Management Obligations for HEALTH *and* SAFETY

Management Obligations for HEALTH *and* SAFETY

GREGORY W. SMITH

CRC Press
Taylor & Francis Group
Boca Raton   London   New York

CRC Press is an imprint of the
Taylor & Francis Group, an **informa** business

CRC Press
Taylor & Francis Group
6000 Broken Sound Parkway NW, Suite 300
Boca Raton, FL 33487-2742

Printed in the United States of America on acid-free paper
Version Date: 2011913

International Standard Book Number: 978-1-4398-6278-0 (Paperback)

Visit the Taylor & Francis Web site at
http://www.taylorandfrancis.com

and the CRC Press Web site at
http://www.crcpress.com

Contents

Preface

> "BP gets it and I get it too—I recognize the need
> for improvement."[*]

Maybe. Maybe not.

The idea for this book was not inspired by any one event. It is the culmination of many, many years of working at first with organisations but more lately with individuals as they have had to come to terms with the consequences of workplace accidents—particularly fatalities.

No one person ever reacts the same to finding him- or herself responsible for managing the aftermath of a death at work, or having to deal with the immediate pressure of being subject to interviews and investigation by safety regulators (much less the drawn-out experience of the legal process), but one of the most constant reactions is *"Why didn't anybody tell me about this?"* It is the same reaction that I get from managers and supervisors whenever I am running training about their individual responsibilities for safety and health in the workplace:

Why didn't anybody tell me about this?

What they are talking about is their personal, individual, and legal responsibilities as a manager for safety and health in the workplace.

Although, as I said, no one event motivated me to sit down and write about individual responsibilities for workplace safety and health, when I started to write this book, oil was continuing to spill into the Gulf of Mexico many months after the catastrophic explosion and fire on board the Deepwater Horizon drilling rig. The political, commercial, and environmental consequences of this event could never be understated, but it should never be forgotten that 11 people lost their lives in the incident—the ultimate price to be paid for a failure of effective safety management in the workplace. It had been only six years since BP suffered an earlier

---

[*] Lord John-Browne, CEO of BP, in a teleconference with reporters shortly after the release of the Baker Panel Review into the causes of the BP Texas City Refinery explosion

catastrophic workplace accident when an explosion at their Texas City refinery killed 17 people and injured over 170, and no doubt the inquiries and investigations will examine the parallels.

The images from the Gulf of Mexico had a special impact on me because I had only just finished my involvement in a frighteningly similar event in Australia (albeit with less disastrous consequences), the uncontrolled escape of hydrocarbons from the West Atlas drilling rig in the Timor Sea and subsequent public inquiry, the Montara Inquiry.*

Much has been written and no doubt will continue to be written about the organisational causes of catastrophic workplace accidents. But when all is said and done those organisations are nothing more than the sum of their parts: people. Organisations do not cut budgets; they do not write safety procedures or processes, or determine training programs. People do. And so it is people who have an impact on safety and health in the workplace.

In 2009 I was presenting at the APPEA conference in Perth, Western Australia, about my views on the minimum expectations that courts and tribunals have of managers and their obligations concerning safety and health. Following the presentation, I was approached by a number of organisations to give similar presentations internally, which ultimately led to a very busy year running workshops on management obligations for safety and health. Routinely I was told that I should write a book on the topic quite simply because none of the participants had ever had their legal obligations, and the relationships between these legal obligations and effective safety management, explained to them—and they felt that there was nowhere that they could go to get this information.

After about 12 months of running training programs, I was involved in a case with an electrical supervisor who was being prosecuted over an incident involving an electrical apprentice. The apprentice had been tasked to do a routine check on a piece of equipment. The equipment had been isolated and de-energized, but the apprentice had gone to the wrong part of the plant and worked on a live piece of machinery. Although he suffered a minor electrical shock, he was not badly injured and was back at work within a few hours. The organisation involved in the incident carried out its normal investigations as part of their routine safety management processes, and the relevant authorities were notified. As far as everyone was concerned the matter was finalised.

Two and a half years later, however, it was the electoral supervisor who was served with a summons and prosecuted.

Ultimately we achieved a good legal outcome for the supervisor, although I am sure that the experience left a mark in its own way.

But during the course of preparing for the trial, he commented to me that although he had been in supervising roles for the best part of 15 years

---

* See http://www.montarainquiry.gov.au/

he had never been exposed to any specific discussion or guidance about his responsibilities as a supervisor. Nothing beyond such clichéd, motherhood statements as "duty of care", "leadership", "leading by example", and the ubiquitous "walking the talk". And he left me with that ringing comment: *"Why didn't anybody tell me about this?"*

My hope for safety managers, line managers, senior executives, chief executive officers, and board members alike is that this book can at least provide a first step to helping them understand the true nature of the expectations that are placed on them by virtue of the obligation to provide a safe workplace.

And if I may, I would also make a plea to the community of safety professionals: just because this is a book written by a lawyer, do not write it off as nothing more than a guide to legal compliance or backside covering. Look at what the cases and inquiries tell us about the expectations of managers. Ask yourself, would the safety culture and safety performance of your organisation be better or worse if your managers, supervisors, and executives understood and met these expectations? But, above all, use the real-life examples provided by the cases to educate and influence those that you are charged with advising, and shift your organisation ever forward to improving safety performance.

The author

Greg Smith has spent almost two decades specialising in occupational health and safety (OHS) within Australian law. His focus is on assisting clients to develop and deliver management obligation programs, particularly to the mining and oil and gas industries.

As a leading OHS practitioner, Greg Smith has deep technical expertise, providing some of Australia's largest and most significant employers with strategic advice on health and safety compliance, incident investigation, management and response, contractor safety management, and representation in health and safety prosecutions. His industry experience is broad: he has applied his OHS expertise to the mining, oil and gas, construction, telecommunications, banking, manufacturing, defence, local and state government, and transport sectors.

From 2007 to 2009, he was employed as the Principal Safety Advisor for Woodside Energy Limited. Reporting to the Vice President for Safety and Health, he was responsible for the ongoing development and implementation of Woodside's global safety management strategy.

In 2010 he acted for a number of parties in the Montara Commission of Inquiry, investigating the blowout and uncontrolled release of hydrocarbons from the West Atlas drilling rig in the Timor Sea off the coast of Western Australia in August 2009.

He has also devised and delivers comprehensive safety and health training programs on behalf of Freehills, a leading Australian-based international commercial law firm, where he is currently a consultant. He also teaches accident prevention at the School of Public Health, at Curtin University in Perth, Western Australia.

Greg Smith graduated from the University of Western Australia in 1990 with a Bachelor of Jurisprudence and a Bachelor of Laws and is admitted to practice in the Supreme Court of Western Australia and the High Court of Australia.

He currently works in Perth, Western Australia.

Introduction

> "You will always have to bear the burden of knowing that Tony Krog died, leaving his wife a widow, because you did not care enough to train him adequately and because you sent him out in a truck when you knew it had inadequate brakes."[*]

On 12 February 1991, Tony Krog was killed when the truck that he was driving ran out of control and overturned. The truck was a second-hand truck and had been delivered to the work site only about a week before the accident. The mechanic who fitted a radio to the truck driven by Mr Krog identified that the brakes were in poor condition and reported that fact to Timothy Nadenbousch.

Mr Krog was employed by Denbo Pty Ltd, a small family construction company run by Timothy and Ian Nadenbousch, who were the only shareholders of the company. At the time of the accident Timothy Nadenbousch was responsible for overseeing the relevant operations, the maintenance of the plant and vehicles, and training drivers and plant operators.

The company was charged with and pleaded guilty to manslaughter. Timothy Nadenbousch was charged with and pleaded guilty to two breaches of the Victorian Occupational Health and Safety Act 1985 in his capacity as an officer of the company, for failing to maintain a safe working environment. He was fined $40,000.

In passing sentence, the Court said:

> The company, as the employer of the deceased, was responsible for ensuring that the condition of the vehicles it required its employees to drive were safe and that the drivers were given proper instructions. The company's work on the project was behind time.

[*] R v Denbo (1994) 6 VIR 157 (Denbo)

Putting the trucks into work was obviously given a higher priority than the safety of the workers.

There was criminal negligence on the part of the company in failing to establish an adequate system for maintenance for its plant and vehicles, in failing to properly train its employees, in permitting [the truck] to be put into use without proper maintenance and in creating a situation where [the truck], with its grossly defective brakes, was capable of being allowed on to the steep track where it was not capable of being kept under control.

Timothy Nadenbousch acknowledged that he had been given and had accepted the responsibility within the company for the maintenance of vehicles and the training of its employees as part of the company's duty to provide and maintain a safe working environment for its employees. However, he did not act responsibly. He was aware of the poor state of the brakes on both trucks but he directed that they may be used. Moreover, the training which he gave the deceased and the other truck drivers was quite inadequate. There was willful neglect in the terms of s52 of the Occupational Health and Safety Act.*

When accidents happen there is a very natural tendency to try to identify someone at fault, and in cases like Denbo, is seems fairly straightforward to be able to assign blame and hold individuals accountable. Clearly no responsible manager should allow workers to operate machinery that he or she knows to be unsafe. The position is not always so straightforward, particularly when it comes to looking at the role of individual managers who are very often quite removed from the day-to-day operations of the businesses that they are responsible for, and who are all too often engaged for their skills and experience in areas unrelated to safety. This dilemma and the changing approaches to it over time are well summed up by Professor Kletz (2001: 5–6):

In fact very few accidents are the result of negligence. Most human errors are the result of a moment's forgetfulness or aberration, the sort of error we all make from time to time. Others are the result of errors of judgment, inadequate training or instruction or inadequate supervision.

* Ibid

Accidents are rarely the fault of a single person. Responsibility is usually spread amongst many people. To quote from an official UK report on safety legislation:

The fact is—and we believe this to be widely recognised—the traditional concepts of the criminal law are not readily applicable to the majority of infringements which arise under this type of legislation. Relatively few offences are clear cut, few arise from reckless indifference to the possibility of causing injury, few can be laid without qualification at the door of a single individual. The typical infringement or combination of infringements arises rather through carelessness, oversight, lack of knowledge or means, inadequate supervision, or sheer inefficiency. In such circumstances the process of prosecution and punishment by the criminal courts is largely an irrelevancy. The real need is for a constructive means of ensuring that practical improvements are made and preventative measures adopted.

In addition, as we shall see, a dozen or more people have opportunities to prevent a typical accident and it is unjust to pick on one of them, often the last and most junior person in the chain, and make him the scapegoat.

The views I have described are broadly in agreement with those of the UK Health and Safety Executive. They prosecute, they say, only 'when employers and others concerned appear deliberately to have disregarded the relevant regulations or where they have been reckless in exposing people to hazard or where there is a record of repeated infringement.' They usually prosecute the company rather than an individual because responsibility is shared by so many individuals.

However, since the earlier editions of this book were published, the advice just quoted has been forgotten and attitudes have hardened. Though penalties have increased there are demands for more severe ones and for individual managers and directors to be held responsible. Many of these demands have come from people and publications that have shown sympathy for thieves, vandals and other lawbreakers. We should understand, they say, the

reasons, such as poverty, deprivation and upbring-
ing that have led them to act wrongly. No such
excuses, however, are made for managers and directors; they just put profit before safety.

The reality is different, as the case histories in
this book will show. Managers and directors are not
supermen and superwomen. They are just like the
rest of us. Like us they fail to see problems, do not
know the best way to act, lack training but do not
realise it, put off jobs until tomorrow and do not do
everything that they intend to do as quickly as they
intend to do it. There are, of course, criminally negligent managers and directors, as in all walks of life,
but they are the minority and more prosecutions
will not solve the real problems.

If anything, the attitudes towards managers and the expectations of
them seem to have hardened further.

Of course, in some cases managers are negligent, even criminally so.

On 25 March 1911, 146 women working at the Triangle Shirtwaist
Factory were killed in a fire that tore through the eighth, ninth, and tenth
floors of the Asch Building in New York City. Most of the women died
from the fire itself or from jumping from the building to escape it. Many
of the women could not escape from the burning building because the
managers would lock the doors to the stairwells and exits to keep the
workers from leaving early. The company's owners were put on trial, but
a jury acquitted them of criminal charges, although they did lose a subsequent civil suit.

Eighty years later, in 1991, 25 workers died in a fire at the Hamlet
chicken processing plant in North Carolina. One of the contributing factors to the number of fatalities was that some of the emergency doors had
been locked shut—history, it seems, is destined to repeat. The owner of the
company, his son (the operations manager), and the plant manager were
all charged with non-negligent manslaughter; there was no trial, however,
as the owner pleaded guilty to 25 counts of voluntary manslaughter while
his son and the other man went free as part of a plea bargain. He received
a prison sentence of 19 years and 11 months but was released just less than
four years into his sentence.

However, more often than not, line management accountability is
less clear.

In September 2002 the yacht Excalibur was sailing down the east coast
of Australia, returning to Victoria after competing in ocean racing events
off the coast of Queensland. On 16 September 2002, while off the coast of
New South Wales, the yacht capsized and four crewmembers died.

The tragedy occurred because the keel of the yacht failed, and the evidence in relevant legal proceedings was that the keel failed because it had been negligently cut and re-welded during the course of its manufacture. The keel had been manufactured by Applied Contract Engineering Pty Ltd, and both the managing director of Applied Engineering, Alex Cittadini, and one of his employees, Adrian Presland, were charged and tried on four counts of manslaughter by criminal negligence. Mr Presland was acquitted, but Mr Cittadini was convicted and initially sentenced to three years in jail almost seven years after the incident.

On 18 December 2009, Mr Cittadini was acquitted.[*]

Essentially the basis upon which Mr Cittadini was convicted was that he had failed to implement reasonable measures to supervise the process of manufacturing the yacht and failed to provide adequate quality control of the process.[†]

The Court identified that a more effective system to verify the quality of the work could have been implemented, but that:

> Although a greater level of supervision of the construction was obviously possible the evidence does not allow of a conclusion that the level of supervision was inadequate much less to the criminal standard.[‡]

The Excalibur case in many ways epitomizes the difficulties that managers can have in coming to terms with their obligations for safety and health. On the one hand, Mr Cittadini had not breached his *"legal"* obligations, at least insofar as criminal charges of manslaughter were concerned. On the other hand, it took the best part of seven years to finally reach that position. And leaving all issues of liability aside, four people died in what was in all likelihood a preventable accident.

The purpose of this book is to help managers understand the impact that their acts and omissions have on the safety and health of the workplaces that they are responsible for by examining the way that courts, tribunals, and inquiries have reviewed, interpreted, and commented on those acts and omissions.

One message that could be taken away from the Excalibur case is that Mr Cittadini was acquitted of a criminal offence. However, in my view, this would be a limited perspective to take, even from a legal liability/legal risk management perspective, because, as the Appeal Court rightly identified:

---

[*] Cittadini v R, R v Cittadini (2009) NSWCCA 302. (Cittadini)
[†] Ibid [4]
[‡] Ibid [80]

> There is no doubt that a more effective system to
> verify the quality of the work done on the yacht
> could have been implemented.*

These types of serious legal processes and findings are not unique to
one jurisdiction, industry, or type of legal proceeding. Following workplace accidents, particularly significant events such as the BP Texas City
refinery explosion, the Esso Longford gas plant explosion, the Montara
uncontrolled hydrocarbon release, and most recently the Deepwater
Horizon event in the Gulf of Mexico, there are typically a series of investigations, inquiries, and legal proceedings directed at trying to determine
what happened—what caused the incident. These processes contribute
greatly to our understanding of organisational elements of a major event,
including issues such as safety culture and organisational learning.

What does not receive as much attention is the role that individual
managers play in the conclusions that are ultimately reached about
organisational failure. And yet the actions and behaviour of managers
and senior executives inescapably create organisational success or failure.
An organisation does not, of itself, do anything; it acts only through the
collective decisions, actions, and omissions of its management and workforce. As has often been quoted, corporations "have neither bodies to be
punished, nor souls to be condemned".†

So, why do we need another book about safety management?

There are many ways to look at safety management and the steps that
need to be taken to provide safe and healthy workplaces. Safety management can be considered in the context of technical or engineering controls
to ensure safety. Another perspective is to consider individual worker
behaviour; what does an organisation need to do to ensure that its workers
"work safely"? Many current commentators examine safety management
through the lens of *"safety culture"*, those organisational characteristics
and traits that improve or diminish safety performance.

In any discussion of safety management or any major accident analysis, you will typically find a combination of all of these elements as part
of the overall picture.

This book looks at safety management from the perspective of management obligations. What expectations are placed on managers at all levels of
an organisation, to ensure that the workplace and systems of work are safe,
and how are these expectations considered and analysed by courts and
public inquiries? As importantly, the book will also consider how manage-

---

* Ibid [78]

† Originally attributed to Edward Thurlow, 1st Baron Thurlow, Lord Chancellor of Great
  Britain from 1778 to 1783, and again from 1783 to 1792

ment actions in relation to these obligations and expectations influence, positively or negatively, the safety performance of an organisation.

All organisations have, as a minimum, legal obligations to ensure that workplaces and systems of work are safe. Many organisations have health and safety aspirations over and above relevant, minimum legal obligations. These aspirational objectives are often reflected in statements such as *"Zero harm"* or *"No one gets hurt. No incidents."*

Whatever the basis for an organisation's safety actions or aspirations, it is generally accepted that the best means of meeting those obligations or aspirations is through a systematic approach: the development of a safety management system. Fundamentally, safety management is built on four core elements. They are:

1. The organisation identifies hazards in their business or undertaking.
2. The organisation assesses the risks arising from those hazards.
3. Controls are developed which are designed to manage the risks to acceptable levels.
4. There is effective supervision to ensure that the controls are properly implemented and effective to manage the risks as intended.

However, a safety management system in and of itself does not ensure safety. It is only effective implementation and oversight of the system that can provide any assurance that safety is being properly attended to in a workplace. Any level of examination of major accidents will reveal safety management systems that failed to control the risks in a business, cases where mandatory procedures were not followed, scheduled hazard-identification processes were not completed in accordance with the requirements of safety management system, or the findings of audits or incident investigations were not closed out as required.

The concept of safety culture will be looked at throughout this book, but at this point it is worth noting that it seems generally accepted that the quality and effectiveness of an organisation's safety management system is determined largely by the organisation's safety culture. So, for example, an organisation's system might require that all incidents be investigated; however, it is the organisation's culture that will determine the quality and value of those investigations. For example:

- Does the organisation have a culture which encourages open and honest feedback about the circumstances of an incident?
- Does the organisation commit adequate resources to incident investigations, including training, personnel, and time?

- Does the organisation call in outside expertise to help in the investigation process?
- Are managers, including the most senior levels of management, open to having their actions examined as part of an incident investigation?
- Is the organisation able to openly share the lessons arising from an incident?

What is also generally accepted is that the main driver of an organisation's safety culture, the factor that influences safety culture the most, is the workforce's perception of management's commitment to safety[*]—in its simplest terms, "*What interests my boss fascinates me.*" If workers really believe that their managers genuinely value safety, then they too will value safety.

What this book seeks to do is to examine the public record of inquiries, investigations, and court cases to understand the expectations that society has of managers, as evidenced by the various legal and quasi-legal processes, for ensuring safe workplaces.

What are individual management obligations for safety and health? As a chief executive officer, chief operating officer, project manager, or general manager, what am I expected to do to ensure the safety and health of my workforce and the technical integrity of my facilities? In the event of a fatality or catastrophic workplace accident, what am I expected to be able to demonstrate and what questions would I be expected to answer?

The philosophy behind this book is, I hope, a simple one. It is founded on the belief that managers want to positively contribute to the safe and healthy operations of their workplaces, but that for the most part they are overwhelmed by the seeming complexity of health and safety management. It seems to me, based on my own beliefs and observations and dealing with organisations and their safety management systems, that no manager comes to work positively disregarding safety and health. None of the managers that I have ever dealt with and nothing that I have seen in any accident enquiry could ever convince me that a manager sought the adverse outcomes of a workplace accident.

This is a book drawn, above all else, from legal and quasi-legal processes. It relies on the strengths and weaknesses of examples of those processes to help individual managers identify and understand their obligations and the behaviours that they are expected to be able to demonstrate that contribute to a safe workplace.

What I hope is one of the more enlightening and thought-provoking aspects of this book is the extensive use of cross examination taken from various proceedings. The examples of cross examination chosen, while designed to reinforce the arguments made in the various chapters where

---

[*] See for example Gadd and Collins (2002) and O'Dea and Flin (2003).

they are used, are not in any way designed to suggest any fault or culpability on the part of any individual manager or their organisation. In most cases the cross examination represents a very small portion of what typically amounts to thousands of pages of evidence.

But the use of cross examination is, in my view, especially relevant to a book about management obligations, if for no other reason than to show how managers' actions are examined in legal proceedings and public inquiries; it gives a realistic and meaningful insight into how management actions and decisions will be challenged and interpreted "after the event". More than that, however, I hope that the use of examples of cross examination illustrates two key points.

First, the universality of the types of questions that managers can expect to be asked. The cases from which the questioning of managers are drawn cover a range of geographical jurisdictions (Australia, the UK, Canada, and the US), and well as different types of inquiries: legal proceedings, health and safety prosecutions, judicial and quasi-judicial inquiries, civil proceedings, and criminal proceedings. Yet in all of these cases the types of questions and the themes underlying these questions are remarkably similar, pointing to a consistent expectation of managers when it comes to discharging obligations for safety and health.

If we can distil and understand these expectations, we can begin to give managers a clearer picture of what they should be doing to meet them, and in turn positively contribute to improved safety outcomes.

A second key point to understand the level of scrutiny that management decisions come under in legal proceedings, a level of scrutiny that, in my experience, management is never subject to in the normal health and safety management processes that exist within organisations. To do justice to safety management systems and to do the right thing by managers and senior executives, they all need to be subject to much greater investigation and scrutiny as part of regular safety management. Far better to be examined in the relative privacy of an organisation and prevent catastrophic events than to be subject to the extraordinarily high level of scrutiny over their role in safety management that occurs in the glare of a public inquiry.

It is hoped that by understanding elements of legal compliance, managers will come to recognise that good safety management and legal compliance are two sides of the same coin. After all, the consequences of failed safety management are far more significant than those contemplated by the legal process.

On 14 November 1996, four workers were killed during underground coal mining operations using a continuous miner at the Gretley Colliery, Wallsend, in New South Wales, Australia. At about 5:30 am a hole opened up at the coalface where they were working, and they were killed by a sudden inrush of water from abandoned workings known as the Young

Wallsend Colliery old workings. In response to the tragedy a coronial inquest and judicial inquiry were conducted, and a range of charges were laid against corporate and individual defendants under New South Wales health and safety legislation. One of those defendants was the mine manager at the time of the accident, Richard Porteous. As part of the sentencing process, he said the following:

> I have always felt the death of Damon Murray, Mark Kaiser, John Hunter and Ted Batterman very keenly and deeply. I know that this will be with me until I breathe my last.
>
> I was the mine manager at Gretley at the time of the accident and so do wear the ultimate responsibility of the death of these poor men. As mine manager I was responsible for the safety and well-being of all persons at the site, so when a person was injured I saw this as a personal failure. With the deaths of four good men this feeling is magnified to incalculable proportions which I've had great difficulty in dealing with. I may never fully deal with it. I know I will be remembered as the mine manager of Gretley regardless of whatever else I have done all do. Everything else pales into insignificance.[*]

References

Gadd, S. and Collins, A. 2002. *Safety culture: A review of the literature*. Sheffield: Health & Safety Laboratory.

Kletz T. 2001. *Learning from accidents*. 3rd ed. Oxford: Gulf Professional Publishing.

O'Dea, A. and Flin, R. 2003. *The role of managerial leadership in determining workplace safety outcomes*. Suffolk: HSE Books.

[*] Statement of Mr Porteous, mine manager at the Gretley coal mine at the time of a fatal inrush in November 1996, quoted in Stephen Finlay McMartin v Newcastle Wallsend Coal Company Pty Limited & Others (2005) NSWIRComm 31, [185] (The Newcastle Wallsend Coal Company)

chapter 1

Managers and safety management

"Having the right intention or attitude to safety is not
enough. The attitude must be backed up by action."[*]

Introduction

The exposures and liability faced by individual managers can vary significantly across industries and jurisdictions. They range from personal prosecutions under health and safety legislation to criminal charges such as manslaughter and gruelling public examination in major inquiries such as those recently seen following the Montara and Deepwater Horizon well blow outs in Australia and the Gulf of Mexico. Regardless of where a disaster occurs, the legal jurisdiction, or the type of proceedings, there is a consistency to the type of questions that managers are asked following serious workplace accidents—questions that are founded on a basic line of inquiry:

What did you do?

Do you know the hazards that arise from the operations that your business undertakes? Do you know the risks associated with those hazards? Have controls been developed to manage those risks? Do you know whether the controls are well implemented and effective to control the risks?

What was the last business decision that you made? Was it consistent with the policies and procedures designed to ensure a safe workplace? Do you know or did you even turn your mind to the question of whether or not that decision affected the safety of the workplace?

How would you demonstrate that you understood and discharged your obligations for safety and health in the workplace?

Society has always recognized that those who create risks have a responsibility to control those risks and are accountable for the consequences when those risks are not controlled. The Code of Hammurabi, an ancient law that dates to about 1760 BC and is on display at the Louvre Museum in Paris, states that if a builder builds a house for someone, and

[*] Silent Vector Pty Ltd v Shepherd & Anor (2003) WASCA 315, [23]. (Silent Vector)

1

does not construct it properly, and the house which he built falls in and kills its owner, then the builder shall be put to death.

Even the Bible recognizes the need for good safety management:

> When you build a new house then you shall make a parapet for your roof, that you shall not bring guilt of bloodshed on your household if anyone falls from it.*

If you type *"safety management"* into Google, it will return something in the order of 1,840,000 hits—at least it did when I tried it. Leaving aside issues of quality and relevance of the moment, it is at least indicative that there is a lot of information potentially available to those interested in the topic of safety management. But having a lot of information available to us does not necessarily help us understand effective safety management or our role as individual managers for safety management.

In this chapter I want to look at two foundation concepts necessary for managers to understand their role in safety management. They are:

1. Safety management systems: The processes and structures in place in an organisation to manage safety, the limitations of safety management systems, and some of the risks in over-reliance on a safety management system.
2. Safety culture: The environment within which an organisation tries to implement its safety management system, the impact that it has on effective safety management, and a manager's role in helping to influence an organisation's safety culture.

These foundation concepts are designed to illustrate some of the complexities of safety management in a way that I hope will encourage managers to challenge current practices for ensuring safety—their own and their organisation's—and demonstrate the need for ongoing assurance that health and safety risks in the business are being effectively controlled.

To start to understand the relationship between these foundation concepts and safety management, consider the following scenario involving two workers working at height.

Two very experienced construction workers were tasked to check and clean bolts that were part of the design of a fan unit to be used in a major industrial building project. The work is being undertaken at a height of more than 30 meters above an open void with an unobstructed fall from the worksite to the ground directly below. The work involved using a 6-kg torque wrench to remove the bolts from one face of the fan unit. The

---

* Deuteronomy 22:8

torque wrench was just less than 2 m long. Each of the bolts that needed to be removed and replaced weighed just over 1 kg. Anything dropped by either of the two workers would represent a very real risk to anybody working below, including the risk that someone could be killed.

The workers, being experienced construction workers, recognized the potential risk of a dropped object—notably one of the bolts—arising from the work that they were undertaking. Having recognized the risk, they implemented controls that they were comfortable with and that they felt controlled those risks. First, they chose to use the "tight" socket to remove the bolts. The socket was so tight that it had to be capped on with a hammer. The second control was for one of the workers to keep his hands cupped underneath the bolt while it is removed—essentially to catch it if it did get dropped.

The job required the removal and replacement of 24 bolts from two fan units, 48 bolts in total. Forty-seven of the bolts presented no problem. One bolt had been cross-threaded, meaning that it could not be replaced until both the bolt and the bolt hole threads had been repaired.

All of the work on the remaining bolts had been completed, and the workers were doing other tasks while the thread was being repaired. The workers were notified that the thread had been repaired about five minutes before the scheduled end of the workday. One of the workers decided that they should go to the worksite and finished the job. The other worker said not to bother, as it was not urgent and they could finish the next morning. The first worker decided to go back to the site and do the job himself. He did so, but while doing the job the bolt slipped and fell the full distance to the ground level, leaving a substantial divit in the concrete floor.

No one was injured by the incident.

From one perspective this incident is simple, and is simply explained: the worker should not have gone and done the job by himself. But within this "near miss" incident are clues, warning signs, of potentially significant safety management system failures. Consider the following:

- The work had originally been scheduled to be done at the ground level before the fan units were lifted into place. A change in work scheduling and the availability of cranes meant that the fan units were put into place without the work having been done. No risk assessment or management of change process was applied even though the safety management system required it.
- The supervisor did not check the way the workers had set up their work or how they were planning to carry out the work even though the safety management system required that to be done.
- The number of supervisors on site that day was less than required under the safety management system.

- The relevant supervisor had not undertaken any of the supervisor training mandated by the safety management system.
- There was a site safety procedure that mandated the following minimum controls when working at heights to manage the risk of dropped objects:
  - All the voids were to be "planked out" to eliminate any void.
  - A barricaded exclusion zone was to be established below the work area, taking into account the potential area within which a dropped object could fall.
  - All tools used at height were to be secured using lanyards.
- Neither the workers nor the supervisors were aware of these specific requirements. There was no evidence that they were enforced or understood on site. The two workers had not been trained on them, and the requirements were not dealt with in the site induction (although 20 minutes of the site induction was dedicated to the use and maintenance of hard hats).
- There were no lanyards on site that were suitable to be used with the torque wrench.
- There were a limited number of suitable torque wrenches on site, and the worker who went back to finish the job had found it difficult to obtain a torque wrench for the work they were doing that day. He was concerned that if he did not complete the job while he still had the torque wrench, he may not be able to easily get hold of the wrench the next day to finish the work.
- Although this worker was an experienced construction worker, he was new to this project.
- If the workers miss the bus back to the accommodation at the end of the scheduled workday, they could have to wait up to half an hour for the next bus if they could not otherwise get a lift.
- The workers work 12-hour shifts in temperatures in the mid- to high 30s Celsius.

Considering these circumstances, there may be more to this incident than simple "operator error" and it may be entirely understandable why the worker did what he did.

The management challenge is to understand what needs to be done to minimise these types of events.

Our understanding of safety management, particularly in the context of both preventing and understanding accidents, has developed over time, moving away from a primary focus on operator error to a more holistic understanding of the overall environment including factors within the organisation as well as external influences such as government regulation. While there is no doubt that human error/operator error is often the starting point for understanding accidents (and in many cases the end point!), the refocusing of attention away from operator error to organisational issues has implications for the expectations placed on managers and their obligations for ensuring a safe workplace.

In most situations it is no longer valid to simply ask what a worker may or may not have done that led to an unsafe situation; it is just as important to understand the role that managers played and the decisions they made or did not make that also contributed to the unsafe situation.

Safety management systems

There is, particularly within large organisations, a tendency to organise the management of health and safety obligations and aspirations through the development and implementation of safety management systems. These are formal, documented, systematic structures designed to identify and manage health and safety risks in the workplace. The structure and content of a safety management system will vary across jurisdictions, industries, and individual businesses, and although the evidence is not conclusive, it does seem to indicate that formal safety management systems can help deliver good safety performance. However, the potential for improved safety performance through a safety management system also carries with it some risk. Safety systems can succeed:

> But in the wrong circumstances they will also fail.
> (Gallagher et al. 2001: vii)

The illusion of safety

Borys (2009) looked at the operation of risk awareness programs in two Australian workplaces. Risk awareness programs are designed to encourage workers to stop and think about risks immediately before commencing work. A typical structure would require a worker to physically step back from the work that he or she is about to undertake and pause for a predetermined period of time; two minutes, five minutes, whatever was mandated. The purpose of the time and space is to give workers an opportunity to think about the task that they are about to undertake and consider whether they had accounted for all of the possible hazards. The process usually requires a worker to complete some paperwork, typically a checklist, with a final acknowledgement that it is safe to proceed with the work.

Borys identifies that such programs can create an "illusion of safety", whereby managers perceive the paperwork as evidence that the workers are considering risk, but in reality there is no assurance that the paperwork is being completed in a meaningful or competent way.

> Workers did not value the program, in particular the paperwork, viewing the paperwork as a means for managers to protect themselves from any threat of prosecution. A typical worker response was: *I can't see anybody who's doing it other than to do the paperwork* or more bluntly: *It's all arse-covering at the end of the day.* Ironically, the collection of the paperwork by managers gave them a sense of security that workers were working safely; a sense that is well summed up by the following comment: *I feel comfortable that they have a process that enables them to have a safe worksite.* (Borys 2009: 6)

There is a clear risk in an unquestioning, unchallenging reliance on safety management systems and safety processes more generally. Not only may they not manage safety risks, they may increase risk, or in some cases positively contribute to an accident.

BP Texas City

On 23 March 2005 one of the worst industrial accidents in the history of the United States occurred when the BP refinery at Texas City exploded during start-up operation. The disaster resulted in 15 deaths and more than 170 injuries.

The U.S. Chemical Safety and Hazard Investigation Board (CSB), an independent federal agency charged with investigating industrial chemical accidents, undertook an accident investigation into the incident, ultimately finding that a string of technical, procedural, leadership, management, and safety culture deficiencies combined to cause the incident (CSB 2007: 25).

As part of their investigation, the CSB recommended that the BP Global Executive Board of Directors commission an independent panel to assess and report on the effectiveness of BP's oversight of safety management and the safety culture of its refineries in North America.

In October 2005, BP announced the formation of the BP U.S. Refineries Independent Safety Review Panel, to be chaired by former Secretary of State James A. Baker III (Baker et al. 2007).

Both the CSB investigation and the Baker Panel review identified a string of failures ranging from operator error to leadership failures at the highest level of the organisation. These failures included failures of the safety management system.

The Baker Panel review in particular was able to point to safety initiatives that actually seemed to adversely impact on safety. One of these was *initiative overload*.

BP's corporate organisation had mandated numerous initiatives to its businesses, including its US refineries, during the last several years. Some of these have been directly related to process safety, such as the integrity management standard, process safety minimum expectations, and engineering technical practices. Some have been more focussed on personal safety, such as the control of work standard and the driving standard, while others relate to non-safety aspects of HSSE, including environmental and other compliance initiatives. Still other initiatives were driven primarily by commercial considerations, such as the separation for sale to a third party of business assets embedded in sites where BP also conducted refining operations. Each successive initiative has required the refineries to develop plans, conduct analysis, commit resources and report on progress. **While each initiative has been well intentioned, collectively they have overloaded refinery management and staff.** BP's corporate organisation has provided the refineries with little guidance on how to prioritise these many initiatives, and the refineries do not receive additional funding to implement each initiative. As a result, senior refinery mangers used phrases such as "initiative overload," "incoming," and "unfunded mandates" to describe what they perceived as an avalanche of programs and endeavours that compete for funding and attention. The ripple effects are then felt throughout the refinery. Many of the hourly workers interviewed at all of the refineries complained that the large number of initiatives and related paperwork contributed to a heavy workload and prevented the workforce from being as focussed on safety and operations as they would like. They also reported that the repeated launch of each successive initiative made it increasingly difficult for the workforce to take any of these initiatives seriously; many interviewees describe this as the "flavour of the month" phenomenon. (Baker et al. 2007: 86)

From the outset it needs to be recognised that safety management systems are simply a tool to help organisations identify hazards and risks in their business and control them. If the system does not achieve those ends, then it needs to be scrutinised and challenged.

Safety management systems need to meet two important criteria— they need to be capable of being understood and implemented by the workforce, and they need to clearly address critical risks in the business.

Longford

Before BP Texas City, a fire and explosion had occurred at the Esso Longford Gas Plant, in Longford, Victoria, Australia. The disaster killed two workers at the gas plant and disrupted gas supplies to the state for over two weeks.

A Royal Commission was established to look at the causes of the disaster and make recommendations about measures to prevent such events in the future (Dawson and Brooks 1999).

The accident happened when operators of the gas plant introduced hot lean oil into a part of the gas plant process that had been shut down for a period of time. During the shutdown, parts of the plant had become cold, with temperatures of some of the vessels used in the process dropping to as low as −48°C. When the hot lean oil was re-introduced into the cold vessel it fractured, releasing hydrocarbon vapour, which ignited.

Like the various inquiries into BP Texas City, the Royal Commission identified failures and weaknesses at all levels of Esso Longford, including failures in the safety management system. Ultimately, reliance on the safety management system in place at Longford for the safe operation of the plant was misplaced in two important respects:

1. The system overall was not understood or implemented by the workforce.
2. The policies and procedures were not effective in controlling the specific risks leading to the accident.

As to the overall system, evidence was given that it was a "world-class system", compliant with the world's best practice. However, whatever the expectations may have been of the safety management system, ultimately the Royal Commission was left in little doubt that those expectations did not translate into practice, noting that *"even the best management system is defective if it is not effectively implemented. The system must be capable of being understood by those expected to implement it"* (Dawson and Brooks 1999: 200).

> Esso's [safety management system], together with all the supporting manuals, comprised a complex management system. It was repetitive, circular, and contained unnecessary cross referencing. Much of

> its language was impenetrable. These characteristics made the system difficult to comprehend both by management and by operations personal.
>
> The Commission gained the distinct impression that there was a tendency for the administration of [the safety management system] to take on a life of its own, divorced from operations in the field. Indeed, it seemed that in some respects, concentration upon the development and maintenance of the system diverted attention from what was actually happening in the practical functioning of the plant at Longford. (Dawson and Brooks 1999: 200)

Moreover, the Royal Commission also noted that there were specific deficiencies in the safety management system, in that it did not address the specific risks that the workers were trying to deal with on the day of the disaster.

In describing the expectations and guidelines for managing safety at Longford and in the Plant Operating Procedures Manual, the Royal Commission observed:

> The [Plant Operating Procedures Manual] did not comply with the guidelines in critical respects. It did not contain any reference to the loss of lean oil flow and contained no procedures to deal with such an event. Nor did it contain any reference to [Gas Plant 1] shutdown or start up procedures or the safe operating temperatures ...
>
> It is difficult to understand why operating procedures dealing with a lean oil absorption plant did not include any reference to the importance of maintaining lean oil flow in the operation of the plant. Plainly that was something that was fundamental. (Dawson and Brooks 1999: 194–195)

That is not to suggest that the answer is simple. Safety management systems for complex operation are almost by necessity complex. They have to take into account a myriad of process, procedures, and an almost infinite combination of circumstances that could arise over the lifetime of operations. This fact alone should make it clear to managers that reliance on a safety management system alone is fraught with dangers and that more is needed to give the safety management system meaning and effectiveness within an organisation.

Montara and Deepwater Horizon

The lessons of BP Texas City and Longford have been widely publicised, and yet the failures identified continue to emerge in major accidents, in particular the challenge of developing effective safety management systems and effectively implementing them.

Between August 2009 and April 2010, two major accident events occurred in the offshore oil and gas industry that attracted worldwide media attention and called into question the level of safety management within the industry.

On 21 August 2009, an uncontrolled release of hydrocarbons occurred at the West Atlas drilling rig operating off the North West coast of Australia. Thankfully there was no loss of life, but hydrocarbons escaped unchecked into the environment for just over ten weeks.

On the evening of 20 April 2010, a failure of well control allowed hydrocarbons to escape from the Macondo well in the Gulf of Mexico. The hydrocarbons engulfed Transocean's Deepwater Horizon drilling rig, and the resultant explosion and fire on the rig killed 11 people and injured 17 others. Hydrocarbons continued to flow from the well for some 87 days, resulting in, reputedly, one of the largest oil spills in US history.

Both incidents have been the subject of major inquiries and ongoing legal proceedings. But it is the observations that have been made about the safety management systems in place at the time and the parallels with both BP Texas City and Esso Longford that serve to illustrate the difficulties inherent in formulating effective management systems in the two key aspects discussed:

1. The system overall, and its ability to be understood and implemented by the workforce.
2. The effectiveness of the policies and procedures to control the specific risks leading to the accident.

Montara

On 21 August 2009, an uncontrolled release of hydrocarbons occurred from the well head platform in the Montara Development Project about 100 km and 150 km from Cartier Island and Ashmore Reef, respectively, off the North West coast of Australia. At the time, operations were being conducted on the West Atlas drilling rig, which had been contracted by PTTEP Australasia (Ashmore Cartier) Pty Ltd (PTTEPAA) to undertake drilling operations.

In late March 2009, one of the wells in the Montara field, HI, had been "suspended"—plugged to prevent hydrocarbons from escaping. The companies working on the field returned in late August 2009, and were preparing the various wells, including HI, for production when the incident occurred.

The Australian Government responded to the incident on the West Atlas drilling rig, which has become known as the Montara incident, by introducing legislation to establish a commission of inquiry, the Montara Commission of Inquiry.* The Montara Commission of Inquiry report was released in late 2010 (Borthwick 2010).

In its finding, the Montara Commission of Inquiry was very critical of the well suspension processes and well construction standards applied in the management of the work.

Other aspects of the Montara incident and the findings of the Montara Inquiry will be discussed throughout this book, but relevant to safety management systems it made findings in terms that could have been lifted directly from the Longford Royal Commission. In terms of the system overall and its ability to be understood and implemented by the work-force, the Montara Commission of Inquiry observed:

> A number of aspects of PTTEPAA's Well Construction Standards were at best ambiguous and open to different interpretations. The fact that a number of PTTEPAA employees and contractors interpreted aspects of the Well Construction Standards differently illustrates the ambiguity and inappropriateness of the Well Construction Standards.

And the effectiveness of the policies and procedures to control the specific risks leading to the accident?

> The Well Construction Standards ... were themselves inadequate. For example, they did not adequately set out how [to] address risks affecting well integrity that arose during drilling, suspension and re-entry of the Montara wells. The ... Well Construction Standards were also of a generic kind and did not adequately address the well control consequences of a batch drilling operation, which involved the derrick spending significant time away from each well and therefore considerable work being undertaken "offline" (which was not always captured). (Borthwick 2010: 9)

* http://www.montarainquiry.gov.au/

In the case of Montara, the system was not understood and specific risks were not adequately addressed. This was a theme that was repeated with terrible consequences less than 12 months later.

Deepwater Horizon

In April 2010 the Deepwater Horizon drilling rig was completing the final stages of drilling and completing the Macondo well in the Gulf of Mexico when the well "blew out", setting in motion a chain reaction of consequences that will be felt for many years to come. The President of the United States, Barrack Obama, appointed a National Commission of Inquiry* to investigate the causes of the blow out. Without going into the specific failures identified by the National Commission, the same failures in the documented safety management system were identified. For example, in making observations on the overall safety management system employed by Transocean:

> If you look at the [Transocean safety] manual you're really impressed by it. It's a safety expert's dream. Everything anybody could ever imagine is in there. ... because if one looks at it everything under the sun is covered. It's hard to see that a particular place somebody saying symptoms about this, if you see that, do this. This is not said by way of criticism. People tried like hell in this manual to get it right but it may be that when time is short there might have been different ways to make clear exactly what should have been done in a short period of time.†

One of the critical processes identified by the National Commission was the conduct of a "negative pressure test", which was found to have been incorrectly run and interpreted.

> The Commission has identified a number of potential factors that may have contributed to the failure to properly conduct and interpret the negative pressure test that night:

* http://www.oilspillcommission.gov/
† Presentation by Fred Bartlett, Chief Counsel to the National Commission of Inquiry on the BP Deepwater Horizon Oil Spill and Offshore Drilling, Meeting 5: November 8–9, 2010 (Washington, D.C.). http://www.oilspillcommission.gov/meeting-5/meeting-details

- First, there was no standard procedure for running or interpreting the test in either MMS regulations or written industry protocols. Indeed, the regulations and standards did not require BP to run a negative-pressure test at all.
- Second, BP and Transocean had no internal procedures for running or interpreting negative-pressure tests, and had not formally trained their personnel in how to do so.
- Third, the BP Macondo team did not provide the Well Site Leaders or rig crew with specific procedures for performing the negative-pressure test at Macondo.
- Fourth, BP did not have in place (or did not enforce) any policy that would have required personnel to call back to shore for a second opinion about confusing data. (Graham et al. 2011: 119)

It is inconsistent with good safety management practices and the obligations and expectations of managers to ensure a safe workplace that safety management systems are put in place without a proper and constant revisiting of their effectiveness. It is also inappropriate to assume such systems are effective without evidence, and, as we have seen, such assumptions can have disastrous consequences.

Clearly, safety management systems in and of themselves are not sufficient to ensure effective safety outcomes. Something more is needed to translate the expectations and guidelines captured in the safety management system into practice and to give them life within the organisation, something to ensure that the system is responsive to the ever-changing hazards and risks in a business and drives the work practices necessary to ensure safe work places.

That something is generally seen as safety culture.

Safety culture

The term *safety culture* first came into the language of safety management in 1987 when it was introduced by the International Nuclear Safety Advisory Group (INSAG) following a post-accident review meeting of the Chernobyl accident (INSAG 1986); and in a subsequent report by the INSAG, safety culture was highlighted as a fundamental management principle (INSAG 1988). It was a term that found more common expression following Lord Cullen's inquiry into the Piper Alpha disaster, an explosion and fire on the Piper Alpha platform in the North Sea on 6 July 1998, which resulted in 165 deaths.

The Piper Alpha inquiry found a litany of management failures that contributed to the disaster, and as Lord Cullen observed:

> It is essential to create a corporate atmosphere or culture in which safety is understood to be, and is accepted as, the number one priority. (Cullen 1990: 300)

In 2010 entering *"safety culture"* into a Google search returned 414,000 hits. The related *"safety climate"* produced 44,700 hits.

This discussion about safety culture is not intend to be a detailed commentary on what safety culture is or is not, or on the distinction between such terms as *safety culture* and *safety climate*; rather it is an attempt to summarise the various observations about a safety culture and its influence on safety performance and then to consider the role of management behaviour in shaping and driving safety culture.

Underlying this discussion are three important themes:

1. Safety culture is recognized as an important driver of effective safety management and good safety performance.
2. It is the attitudes, acts, and omissions of managers at all levels of an organisation that create the organisation's safety culture.
3. It has become generally accepted that a safety culture and management's role in creating an effective (or otherwise) safety culture is a legitimate avenue of investigation following a workplace accident. In many cases the language of safety culture has become the language of legal proceedings.

Following on from their initial reports about Chernobyl and safety culture, INSAG published further guidance on safety culture, which defined safety culture as:

> that assembly of characteristics and attitudes in organisations and individuals which establishes that, as an overriding priority, nuclear plant safety issues receive the attention warranted by their significance. (INSAG 1991: 4)

The report also identified the requirements of managers:

> The attitudes of individuals are greatly influenced by their working environment. The key to an effective Safety Culture in individuals is found in the

> practices moulding the environment and fostering
> attitudes conducive to safety. It is **the responsibil-
> ity of managers** to institute such practices in accor-
> dance with their organisation's safety policy and
> objectives. (INSAG 1991: 6) [emphasis added]

The importance of managers' influence on safety culture was
described by Carolyn Merritt, the Chairman and CEO of the U.S.
Chemical Safety and Hazard Investigation Board at the time of the Texas
City Refinery disaster:

> A good safety culture is the embodiment of effec-
> tive programs, decision making and accountability
> at all levels. It is a much different concept from sim-
> ply having good procedures on paper.
>   There is a widespread misperception that safety
> culture can be improved solely through modifying
> unsafe worker behaviors. While human errors con-
> tribute to most major incidents including this one,
> they are rarely the root cause. The mistakes that
> were made in Texas City have their roots in deci-
> sions made by managers at the facility and the cor-
> porate level, sometimes years earlier.
>   Thus **when we talk about safety culture, we
> are talking first and foremost about how mana-
> gerial decisions are made**, about the incentives
> and disincentives within an organization for pro-
> moting safety. Are production and cost control
> being rewarded at the expense of safety and risk
> management?
>   One thing I have often observed is that there
> is a great gap between what executives believe to
> be the safety culture of an organization and what
> it actually is on the ground. Almost every execu-
> tive believes he or she is conveying a message that
> safety is number one. But it is not always so in reali-
> ty.* [emphasis added]

What is clear since at least Chernobyl is that inquiries into major accident are concerned with understanding the safety culture of an organisa-

---

* Statement for the BP Independent Safety Review Panel 10 November 2005. http://www.
csb.gov/assets/document/Carolyn_Statement_3.pdf (accessed 23 November 2010)

tion and understanding the role that management played in creating and embedding that culture.

The central importance of safety culture has been evidenced most recently in the findings emerging from the Deepwater Horizon event in April 2010:

> The most significant failure at Macondo—and the clear root cause of the blowout—**was a failure of industry management**. Most, if not all, of the failures at Macondo can be traced back to underlying failures of management and communication. Better management of decisionmaking processes within BP and other companies, better communication within and between BP and its contractors, and effective training of key engineering and rig personnel would have prevented the Macondo incident. BP and other operators must have effective systems in place for integrating the various corporate cultures, internal procedures, and decisionmaking protocols of the many different contractors involved in drilling a deepwater well. (Graham et al. 2011: 122)
>
> …
>
> It is also critical … **that companies implement and maintain a pervasive top-down safety culture** … that reward employees and contractors who take action when there is a safety concern even though such action costs the company time and money. (Graham et al. 2011: 126)
>
> …
>
> The record shows that without effective government oversight, the offshore oil and gas industry will not adequately reduce the risk of accidents, nor prepare effectively to respond in emergencies. However, government oversight, alone, cannot reduce those risks to the full extent possible. Government oversight … must be accompanied by the oil and gas industry's internal reinvention: **sweeping reforms that accomplish no less than a fundamental transformation of its safety culture**. Only through such a demonstrated transformation will industry—in the aftermath of the Deepwater Horizon disaster—truly earn the privilege of access

to the nation's energy resources located on federal properties. (Graham et al. 2011: 217)

...

Offshore oil and gas exploration and production are risky. But even the most inherently risky industry can be made much safer, given the right incentives and disciplined systems, sustained by committed leadership and effective training. **The critical common element is an unwavering commitment to safety at the top of an organization: the CEO and board of directors must create the culture and establish the conditions under which everyone in a company shares responsibility for maintaining a relentless focus on preventing accidents**. Likewise, for the entire industry, leadership needs to come from the CEOs collectively, who can apply pressure on their peers to enhance performance. (Graham et al. 2011: 218)

...

What the men and women who worked on Macondo lacked—and what every drilling operation requires—was a culture of leadership responsibility. In remote offshore environments, individuals must take personal ownership of safety issues with a single-minded determination to ask questions and pursue advice until they are certain they get it right. (Bartilt, Sankar and Grimsley 2011: xi) [emphasis added]

The language of safety culture, it seems, has indeed become the language of legal proceedings.

What then becomes critical is that managers need to understand that they play a primary, central role in developing and driving an organisation's safety culture, and that it is the workforce's perception of management's commitment to safety, their attitudes and behaviour, that provides the best measure of that culture—ultimately it is management's commitment to safety that influences a worker's behaviour.[*]

The various guidance material, literature, and publications looking at safety culture emphasise the importance of leadership and management commitment to safety—visible commitment by management—as being

---

[*] See for example Gadd and Collins 2002 and O'Dea and Flin 2003.

key to developing a good safety culture. However, the content of *"leadership"* and *"management commitment"* can vary tremendously amongst individuals: One person's visible commitment is another's micromanagement.

One of the drivers behind the study of the various cases and inquiries set out in this book is to look for consistent messages about the expectations of managers and the *content* of *"leadership"* and *"management commitment"* in a safety context. From the various studies into safety culture and the role of managers, there are a number of key themes that can be identified as being important, for example:

1. Managers planning work effectively to take account of safety
2. Managers being actively involved in monitoring safety performance
3. Managers being involved in reactively monitoring safety performance—investigating and challenging when things go wrong
4. Managers being visible in safety and health—participating in safety and health committees (Gadd and Collins 2002: 25)

These are themes that continually reappear in investigations into major accidents.

Within these broad behaviours we can find evidence of management decisions and actions (or inaction) that contribute to both safety culture and safety performance. These include such things as developing remuneration and bonus strategies, setting and then driving safety and health policies, site visits and engaging in conversations with workers, reporting and developing processes to respond to safety and health concerns or breaches of expectations about safety and health, and a range of other actions.

Through the cases and various studies we can start to construct a picture of both the expectations placed on managers for safety and health and the steps that managers can take to meet those expectations.

References

Baker, J. et al. 2007. *The report of the BP U.S. Refineries Independent Safety Review Panel*. U.S. Chemical Safety and Hazard Investigation Board, Washington. http://www.bp.com/liveassets/bp_internet/globalbp/globalbp_uk_english/reports_and_publications/presentations/STAGING/local_assets/pdf/Baker_panel_report.pdf.

Bartlit, F., S. Sankar, and S. Grimsley. 2011. Chief Counsel's Report. *National Commission on the BP Deepwater Horizon Oil Spill and Offshore Drilling*. http://www.oilspillcommission.gov/sites/default/files/documents/C21462-407_CCR_for_print_0.pdf (accessed 17 February 2011).

Borys, D. 2009. Exploring risk-awareness as a cultural approach to safety: Exposing the gap between work as imagined and work as actually performed. *Safety Science Monitor*, Issue 2, Volume 13. http://ssmon.chb.kth.se/vol13/issue2/3_Borys.pdf (accessed 23 November 2010).

Borthwick, D. 2010. *The report of the Montara Commission of Inquiry.* Montara Commission of Inquiry, Canberra. http://www.ret.gov.au/Department/Documents/MIR/Montara-Report.pdf. (accessed 25 November 2010).

CSB (U.S. Chemical Safety and Hazard Investigation Board). 2007. *Investigation report: Refinery explosion and fire.* Washington. http://www.csb.gov/assets/document/CSBFinalReportBP.pdf (accessed 23 November 2010).

Cullen, Lord. 1990. *The public inquiry into the Piper Alpha disaster.* Department of Energy. London: HMSO.

Dawson, D. and B. Brooks. 1999. *Report of the Longford Royal Commission: The Esso Longford Gas Plant accident.* Melbourne: Government Printer for the State of Victoria.

Gadd, S. and Collins, A. 2002. *Safety culture: A review of the literature.* Health & Safety Laboratory. Sheffield.

Gallagher, C., E. Underhill, and M. Rimmer M. 2001. *Occupational health and safety management systems: A review of their effectiveness in securing healthy and safe workplaces.* National Occupational Health and Safety Commission. Sydney. http://www.safeworkaustralia.gov.au/AboutSafeWorkAustralia/WhatWeDo/Publications/Documents/127/OHSManagementSystems_ReviewOfEffectiveness_NOHSC_2001_ArchivePDF.pdf (accessed 23 November 2010).

Graham, B. et. al. 2011. *Deep water: The Gulf oil disaster and the future of offshore drilling.* Report to the President. National Commission on the BP Deepwater Horizon Oil Spill and Offshore Drilling. http://www.oilspillcommission.gov/final-report (accessed 11 January 2011).

International Nuclear Safety Advisory Group. 1986. *Summary report on the Post-Accident Review Meeting on the Chernobyl accident.* Safety Series No. 75-INSAG-1. International Atomic Energy Agency, Vienna.

International Nuclear Safety Advisory Group. 1988. *Basic safety principles for nuclear power plants.* Safety Series No. 75-INSAG-1. International Atomic Energy Agency, Vienna.

International Nuclear Safety Advisory Group. 1991. *Safety culture.* Safety Series No. 75-INSAG-4. International Atomic Energy Agency, Vienna.

O'Dea, A. and R. Flin. 2003. *The role of managerial leadership in determining workplace safety outcomes.* Suffolk: HSE Books.

chapter 2

Bata Industries

> "No proper system to prevent the discharge was in place, and reasonable steps were not taken to ensure that the system that was in place was effectively operated ..."[*]

Introduction

On first reading, the Bata Industries case may seem like a strange choice to include in a collection of case studies about management obligations for safety and health, given that it is a Canadian decision dealing with breaches of environmental legislation. But the case is a very valuable study of how individual management behaviour—interpreted within the same management system—can be very different and have different consequences. The case is a good demonstration of how the management system by itself adds little to the effective management of risk. Rather, it is the actions and inactions of individuals within the system, most particularly individual managers, that determine the quality and effectiveness of the system.

In the Bata decision three managers were charged in relation to the same incident, and based on their different positions, acts, and omissions, different results arose.

Background

Bata Industries Limited operated a shoe manufacturing plant in Ontario, Canada. The relevant structures of the entities involved in the case were described by the Court as follows.

The Bata shoe organization comprises some 80 companies around the world. Thomas G. Bata is the chief executive officer. The one company, which is located in Canada, headquartered at Toronto, is Bata Industries Limited. The company has four divisions. Each division operates autonomously under the general manager, and each general manager is a vice-president and director of Bata Industries Limited. The president, also a director, of Bata Industries Limited during the material time was Douglas

---

[*] R v Bata Industries Ltd 7 C.E.L.R. (N.S.) 245, 9 O.R. (3d) 329 (Bata)

Marchant. Thomas G. Bata, who functioned chiefly in an advisory capacity, was chairman of the board and a director of Bata Industries Limited.

The division of Bata Industries Limited that this case involves was the shoe manufacturing division located at Batawa, Ontario. The general manager/director/vice-president on site was Keith Weston.*

Chemical waste from the manufacturing process at Batawa was stored in a large number of various types of containers at the site. Investigations by the Ontario Ministry for the Environment in August 1989 identified a range of problems with the storage of the chemical waste, problems that included:

- Chemical waste containers that were decaying
- Chemical waste containers that were uncovered and exposed to the environment
- Material that was leaking and foaming from the chemical waste containers, including material that was "spraying" from the bung holes of some containers

There was also evidence of staining on the ground, apparently from the chemical waste. Evidence was also presented by a hydrologist that if the chemical waste were spilled, it would migrate into the groundwater and once it entered the groundwater would break down into both known and suspected carcinogens.

The company, and three company officers, Thomas Bata, Mr Marchant, and Mr Weston, were all charged with environmental offences relating to the storage of the chemical waste. Ultimately the company, Mr Weston, and Mr Marchant were all convicted of offences. In the case of Mr Marchant and Mr Weston it was on the basis that they failed to take reasonable care to prevent the discharge.

Thomas Bata was acquitted.

Evidence in the court proceedings established that the storage of the drums and their disposal had been a matter of concern to a number of personnel since 1983. Various company employees, including health and safety committee members, had expressed concerns about the containers and their contents. In 1986 representatives of Bata met with Tricel, a waste disposal company, but it was not until April 1989 that Bata had obtained a quote from Tricel to dispose of the waste.

On 11 August 1989, Tricel was told by Bata to *"get the waste out as soon as possible"*, although the court found that the *"motivating force"* for this direction was the investigation by the Ministry for the Environment.†

---

* Bata, [149–150]
† Bata, [50]

An important theme in the Bata decision is that some managers can be found to have met their obligations even though other managers operating within the same system have not met theirs. The discussion of the various management actions and inactions looks at a number of important concepts that arise in a variety of discussions about management obligations and that are all evident to various degrees in all of the case studies discussed in this book. Those concepts are:

- The extent to which managers understand their responsibilities
- The extent to which managers understand the risks that they are responsible for
- Whether or not managers are aware when risks are not being managed, and the processes that are in place to bring any failure in the way that risks are managed to their attention
- How managers respond when they are made aware that risks are not being properly managed
- What steps managers take to actively and personally understand that risks in their area of responsibility are being managed
- The extent to which managers can delegate responsibility for risk, and the assurances that need to be in place to be satisfied that the delegate is effectively managing the risks

The Court looked expressly at how different managers responded to various systems failures, as well as the specific failure arising from the storage of the chemicals on site, as part of a broader consideration of whether the managers had met the expectations of their roles—specifically whether or not they had exercised all due diligence.

Thomas Bata

Mr Bata was the director whom the Court identified as having had the least personal contact with the relevant site, and in finding that he had discharged his obligations the Court made a number of helpful observations:

- Mr Bata was aware of his environmental responsibilities.[*]

On 4 July 1986, the Bata organisation issued a Technical Advisory Circular (TAC 298) to its various businesses around the world. TAC 298 provided an update on industrial safety issues and had a focus on environmental issues; the first area it recommended for actioning was an assessment of environmental exposures and the reduction of potential risks.

[*] Bata, [66, 153, and 157]

- Mr Bata had processes in place whereby environmental concerns were brought to his attention, and he responded to those concerns when they were brought to his attention.*
- Mr Bata exercised a level of personal verification to satisfy himself that the organisation worldwide was being operated in accordance with the organisational expectations. The Court found that Mr Bata attended the Batawa site once or twice a year, and when he did so, he was a "walk-around director" whose visits could not be orchestrated. Mr Riden, the manager who replaced Mr Weston, gave evidence that "You never knew where Mr Bata was going to go, believe me. He had a habit of trying to outguess where you wanted him to go."†

Ultimately the court found:

> In short, he was aware of his environmental responsibilities, and had written directions to that effect in TAC 298. He did personally review the operation manuals on site, and did not allow himself to be wilfully blind or orchestrated in his movements. He responded to the matters that were brought to his attention promptly and appropriately.‡

Although the discussion about Mr Bata is a good illustration of the steps that senior executives need to be able to demonstrate they took in discharging their obligations to manage risk, dealing with events that occurred in 1989 and the collective evidence of cases and inquires since that time would suggest that further action may be required, judged in the light of current social expectations. There are two issues in particular that require further examination:

- The organisational remuneration strategies and how that may have impacted the management of risk
- Mr Bata's reliance on the system

Remuneration strategies

There was some suggestion by the Court in a passing comment made while looking at whether or not Mr Weston met his obligations that his failure to address the chemical storage issue may have been driven in part

* Bata, [157].
† Bata, [155].
‡ Bata, [154–158]

by his remuneration. The evidence showed that forty percent of his remuneration package was based on incentive. It was a reducing percentage,
decreasing as losses were cut and designed to have a stable company in
the long run.[*]
    And the Court observed:

> When he was transferred in November, he allo-
> cated $100,000 to waste disposal, again without
> further knowledge. One cannot help but wonder if
> his diminished incentive package was a motivating
> factor in the allotment of $100,000 at this time. This
> expense would only affect him personally to the
> amount of $500 because the company was now in
> a profit position and his salary incentive based on
> reducing loss was minimal.[†]

    It is generally recognised and accepted that remuneration strategies
can impact the management of risks, and that remunerations strategies
need to be carefully considered in terms of their potential impact on risk
management, including health and safety management. An independent
report commissioned by the New South Wales Mine Safety Advisory
Council identified:

> Production bonus and safety incentive schemes that
> involve payment in exchange for achieving particu-
> lar outcome targets have not proved themselves to
> consistently or reliably improve safety outcomes.
> The confusion about the presence of such schemes
> evident in our interviews and in questionnaire
> responses suggests that any positive effects are
> likely to be limited at best.
>   The most commonly cited benefit was that the
> schemes encourage effective injury management.
> Rather than a benefit, this could well be seen as a
> cost of the schemes, since responding promptly to
> injuries is a fundamental building block of effec-
> tive OHS financial benefit available, the basics of
> OHS management may not be in place. As the
> Future Inquiry Workshop participants identified,
> in a world class OHS system, people contribute to

[*] Bata, [63]
[†] Bata, [171]

OHS management, not because of extra money, but because it is "the right thing to do".

Generally, sites reported that safety incentive schemes making payments as a result of achievement of outcome targets either made no difference at all or had negative effects on incident reporting. The questionnaire responses suggest that this is more likely where large payments are involved, which further reinforces the negative consequences that may have been realised. (Shaw et al. 2007: vii–ix)

We also see in the BP Texas City refinery explosion criticisms of remuneration and reward strategies that drove management attention and action towards managing personal safety issues rather than process-safety risks,* which, ultimately, were at the heart of the explosion. An investigation conducted by the U.S. Chemical Safety and Hazard Investigation Board (CSB) concluded:

> BP Group implemented an incentive program based on performance metrics, the Variable Pay Plan (VPP), which was in place at the Texas City refinery for several years prior to the ISOM incident. Payouts under the VPP were approved by the refining executive managers in London. "Cost leadership" categories accounted for 50 percent and safety metrics for 10 percent of the total bonus. For the 2003–2004 period, the single safety metric for the VPP bonus was the OSHA Recordable Injury Rate. (CSB 2007: 153)

And

> Beginning in 2002, BP Group and Texas City managers received numerous warning signals about a possible major catastrophe at Texas City. In particular, managers received warnings about serious deficiencies regarding the mechanical integrity of aging equipment, process safety, and the negative safety impacts of budget cuts and production pressures. However, BP Group oversight and Texas City

---

* Personal safety issues generally refer to higher likelihood, lower consequence type incidents, generally framed as "slips, trips and falls". Process safety risks refer to safety risks that have the potential to cause major accident events, such as fires and explosions.

management focused on personal safety rather than on process safety and preventing catastrophic incidents. Financial and personal safety metrics largely drove BP Group and Texas City performance, to the point that BP managers increased performance site bonuses even in the face of the three fatalities in 2004. Except for the 1,000 day goals, site business contracts, manager performance contracts, and VPP bonus metrics were unchanged as a result of the 2004 fatalities. (CSB 2007: 177–178)

It is likely that in today's "legal" environment, particularly in the investigation of major accidents, that similar remuneration strategies could be used to direct criticism at the organisation generally and the managers responsible for developing and approving the remuneration strategies.

Reliance on "the system"

In finding that Mr Bata had met his obligations the Court commented:

"He was entitled to rely upon his system, as evidenced by TAC 298, *until he became aware the system was defective".*

This observation needs to be considered in light of the other findings by the Court that:

- Problems were brought to his attention
- Mr Bata did not allow himself to be "wilfully blind" to risks in his business

Various cases and inquiries have made it clear that organisations and managers are not entitled to simply "set and forget" safety management systems, and we see for example in both the Montara and Deepwater Horizon incidents managers being variously criticised for being *"insufficiently vigilant"* or *"not having a clue about what was going on"* when it came to managing what were seen to be critical health and safety risks in the business.

The current-day expectations for managers is that they will do more than put in place systems to manage risks: managers need to be able to demonstrate what steps they take to understand that the system is implemented and is effective in managing the risks that it is designed to manage.

---

* Bata, [157]

Douglas Marchant

As the Court described it, Mr Marchant presented "another variation in director's liability".[*] Mr Marchant's liability was in the end simply stated: he failed to respond effectively to a problem that was brought to his attention.

The evidence established that Mr Marchant was appointed to the board as president on 26 January 1988, that he attended the Batawa site about once a month, that these visits included a tour of the plant, and that the problem of the storage of the chemical waste was brought to his personal attention around 15 February 1989. In these circumstances the Court found:

> The evidence therefore establishes that for at least the last six months of the time alleged in the charges (February 15, 1989, August 31, 1989), he had personal knowledge. There is no evidence that he took any steps after having knowledge to view the site and assess the problem. There is no evidence that the system of storage was made safer or temporary steps undertaken for containment, until such time as removal could be effected.[†]

It was the lack of effective response by Mr Marchant that ultimately led to a finding that he had not met his responsibilities:

> In the circumstances, it is my opinion that due diligence requires him to exercise a degree of supervision and control that "demonstrate that he was exhorting those whom he may normally be expected to influence or control to an accepted standard of behaviour."
>
> He had a responsibility not only to give instructions, but also to see to it that those instructions were carried out in order to minimize the damage.[‡]

Keith Weston

Mr Weston was employed by Bata Industries from 1984 until November 1988, when he was transferred to Malaysia. This meant that he was not on site for at least eight months prior to the environmental offences being identified.

[*] Bata, [160]
[†] Bata, [161–162]
[‡] Bata, [164–165]

The Court found that Mr Weston did not take all reasonable steps to prevent the environmental offences from occurring, and the overall tone of the evidence was that he did not regard environmental issues as a priority—indeed he gave evidence that environmental concerns were not his first priority and that responsibility for environmental matters were shifted to *"the department responsible for the task"*, which turned out to be an individual, Mr De Bruyn.[*] The following from the case illustrates the approach taken by Mr Weston and the Court's view of it:

> Mr de Bruyn discussed the chemical waste drum storage problem with Mr Weston after TAC 298 arrived. Mr De Bruyn was instructed to get a quote for the cost of removing the waste. In late 1986 or early 1987, Mr Weston was advised that the quote was $56,000. His reaction: "I was extremely surprised. I felt it was a large sum of money. It's an area of the business that I know absolutely nothing about, so I had no way of knowing if it was high or considerably too high. So I instructed Mr De Bruyn to get me an alternative quote."

On cross examination

Q: Did you ever speak to anyone personally at Tricel?

A: No.

Q: And he felt the quote was too high?

A: I felt that I needed an alternative, I didn't have sufficient information to accept that quote per se, I needed an odd turn of quote to assess its value.

Q: ... Surely you must have looked at some writings, some memos or letters from Tricel, didn't you?

A: No, I was informed by Mr De Bruyn that was the quotation that had been given.

Q: And you felt that was too high?

A: I felt it would be irresponsible to spend money without getting an alternative quote.

Q: Between late 1986 and late 1987 ... did you ever think of phoning up the Ministry of Environment and asking them what should be done with the waste?

A: Personally, I did not, no.

Q: Did you ever contact anyone who might have knowledge about environmental matters to give you some advice? By that, I mean a consultant?

[*] Bata, [64]

A: No ... I passed the, the problem was in the hands of the safety and environmental officer, and he was dealing with it.

Q: And the safety and environmental officer was Mr De Bruyn?

A: Yes.

Q: And wasn't it your understanding he certainly didn't have any environmental training, did he?

A: Specific training from environmental control—no[*]

The Court noted that Mr De Bruyn's responsibilities at the time included

- Personnel and property management
- Public relations
- Environmental concerns
- Purchasing officer
- Health and safety committee
- The 50th-anniversary festivities
- The water treatment plant[†]

The Court ultimately found:

> that Keith Weston cannot shelter behind the advice he received from Mr De Bruyn. As Bata was "cut to the bone" by Mr Weston, the additional responsibilities fell upon Mr De Bruyn and grossly overloaded him. The problem was aggravated by the inference from the evidence that Mr De Bruyn was not given authority to expend the $58,000 or $28,000 on his own. He required the approval of Mr Weston.
>
> ...
>
> He had an obligation if he decided to delegate responsibility, to ensure that the delegate received the training necessary for the job, and to receive detailed reports from that delegate.[‡]

While Mr Weston was not prepared to accept a $56,000 quote to clean up the chemical waste, he did accept the alternative $28,000 quote obtained by Mr De Bruyn in late 1987 and on the basis of that quote directed that the waste removal be undertaken, but by mid 1988 the contractor who

[*] Bata, [68]
[†] Bata, [64].
[‡] Bata, [172–173]

provided the lower quote advised that they could not do the work. The Court noted:

> It is my opinion, red flags should have been raised in [Mr Weston's] environmental consciousness when the first quote of $58,000 was obtained. Instead of simply dismissing it out of hand, he should have inquired why it was so high and investigated the problem. I find that he had no qualms about accepting the second quote of $28,000, and he had no further information other than it was cheaper.[*]

The Court found that Mr Weston had experience in the production side of the business and was aware of the toxic chemicals used in the processes, and that he had also been reminded of his responsibilities by TAC 298:

> As the "on-site" director, Mr Weston had a responsibility in this type of industry to personally inspect on a regular basis, i.e., "walkabout". To simply look at the site "not too closely" 20 times over his four-year tenure does not meet the mark.[†]

Mr Weston's actions were also compared to those of two other managers; Mr Riden,[‡] a contemporary, and Mr Richer, Mr Weston's replacement.

Mr Richer had been employed as the vice-president and director of Bata Engineering, which was located next to the site Mr Weston was responsible for. When the assets of Bata Engineering were sold and he became the vice-president and director of Invar, a fence was erected between Bata and Invar.

In relation to Mr Riden's conduct, the Court found that Mr Riden:

- had identified environmental problems associated with oil-based cutting fluids at Bata Engineering

---

[*] Bata, [170]

[†] Bata, [173]

[‡] The fact that Mr Riden, a contemporary of Mr Weston, brought environmental issues to Mr Bata's attention was a fact relied upon by the Court in finding that Mr Bata had met his environmental obligations. In effect, the fact that Mr Bata had one manager who brought environmental issues to his attention meant that he could assume that his other "experienced" manager in the same position in another part of the business would also bring similar matters to his attention; in circumstances where Mr Bata was not being wilfully blind, there was nothing to put Mr Bata on notice that Mr Weston was not bringing, or would not bring, similar matters to his attention, as Mr Riden had been doing.

- discussed his concerns with Mr Bata and was encouraged to proceed with a solution
- developed and implemented solutions[*]

The evidence showed that Mr Riden had no difficulty[†] getting the relevant approvals from Mr Bata.[‡]

When comparing Mr Weston's management of delegation to Mr Riden's, the Court observed that Mr Riden indicated that when the environment became a "hot topic", he would appoint a person to assume environmental responsibilities. He instructed him to attend courses, and to read and follow the procedures. The person was to examine "the rules" and tell Mr Riden what he was doing in order to comply with "the rules".

Ultimately the Court found that Mr Weston's conduct stood in "sad comparison"[§] to Mr Riden's.

The relevant conduct of Mr Weston was very shortly stated by the Court:

> [Mr Weston's] successor, Mr J. P. Richer, was employed by Bata from January 1989 to November 6, 1990. On or about February 15, he toured the plant and saw the barrel storage site. He noticed they were deteriorating badly, some rusting, some open, and there was a chemical smell. He immediately spoke to Mr De Bruyn and notified his superior, Mr Marchant. In his own words, "When I discovered what was on the site, it definitely became a priority for me."[¶]

This conduct stands in contrast to the way that Mr Weston responded to a known problem.

Final comments

Although this case was conducted and needs to be considered in the specific context of the legal processes and legislation that governed environmental offences in Canada in the late 1980s and early 1990s, the principles identified by the Court are common principles that are described and applied in cases and inquiries that examine major incidents in a range of

---

[*] Bata, [73–75]
[†] Bata, [76]
[‡] Bata, [76]
[§] Bata, [172]
[¶] Bata, [54]

industries and jurisdictions today—most recently following the Montara and Deepwater Horizon incidents, already referred to.

Some key messages coming out of the Bata decision that managers need to apply to their own management of workplace health and safety risks are:

- Do I understand what my responsibilities for health and safety are generally, under applicable legislation and as set out in company documents, policies, and procedures?
- Do I know what the relevant health and safety risks are in my area of responsibility?
- Do I positively take steps to review the work areas and risks that I am responsible for to ensure that they are being adequately controlled?
- If I delegate responsibility for health and safety matters, do I know that the people I delegate matters to have the skills, resources and experience to do what I have asked? What reports do I get from my delegates, and how do I follow them up to ensure that health and safety matters are being properly dealt with?

In the context of Mr Weston's focus on his business priorities and the influence this seems to have had on the management of his environmental responsibilities, it may also be worthwhile taking some time to consider whether your business priorities, or the business priorities that you set for others, have any impact on the effective management of workplace health and safety—and how you get assurance that they do not.

As we will see in the cases discussed later in the book, the pursuit of business objectives—sometimes characterised in investigations as "cost vs. safety" or "production before safety"—while seldom a positive, knowing strategy pursued by management, is nonetheless a regular critical factor in major workplace accidents, and managers need to be able to demonstrate that they understand this risk and have assurance that it is being effectively controlled.

References

CSB (U.S. Chemical Safety and Hazard Investigation Board). 2007. *Investigation report: Refinery explosion and fire.* Washington. http://www.csb.gov/assets/document/CSBFinalReportBP.pdf (accessed 23 November 2010).

Shaw, A. et al. 2007. *Digging deeper.* New South Wales Department of Primary Industries.   http://www.dpi.nsw.gov.au/minerals/safety/consultation/wran-consultancy-project-2007/Digging-Deeper-final-report---volume-1.pdf (accessed 29 September 2010).

chapter 3

Management line of sight

> "And now we are all paying the consequences because those of you at the top don't seem to have a clue about what was going on on this rig."[*]

Introduction

Management line of sight is a concept that goes to a manager's understanding of the risks in their business and the level of assurance that those risks are controlled.

If the gap between the system as it is believed to work and the system as it actually operates represents a fundamental misunderstanding of the effectiveness of an organisation's safety management system, line of sight is about trying to understand that gap and close it. What are managers expected to know about the critical risks in their business and how they are controlled? And how do managers "see into" their organisations to gain an understanding of those risks and their controls?

In this chapter I will consider how the concept of management line of sight arises during an inquiry into a major accident, and the types of questions managers are expected to be able to answer.

BP Texas City

As we have already seen in Chapter 1, following the fire and explosion at BP's oil refinery in Texas City in March 2005 a number of major investigations were undertaken into the cause of the disaster and management's role in it.

At the same time as these investigations were proceeding, private legal actions were being initiated on behalf of families who lost loved ones in the disaster.

The combination of these inquiries and proceedings provide a number of different insights into management's line of sight into the risks, and the failure to control those risks, that emerged at Texas City.

Organisationally, there was a failure by BP to properly oversee the risks in its operations at Texas City, and the CSB found:

[*] Hayward 2010: 130

> The BP Board of Directors did not provide effective oversight of BP's safety culture and major accident prevention programs. The Board did not have a member responsible for assessing and verifying the performance of BP's major accident hazard prevention programs. (CSB 2007: 25)

The Baker Panel review made similar observations:

> The Panel believes that leadership from the top of the company, starting with the Board and going down, is essential. In the Panel's opinion, it is imperative that BP's leadership set the process safety "tone at the top" of the organization and establish appropriate expectations regarding process safety performance. Based on its review, the Panel believes that BP has not provided effective process safety leadership and has not adequately established process safety as a core value across all its five U.S. refineries. While BP has an aspirational goal of "no accidents, no harm to people," BP has not provided effective leadership in making certain its management and U.S. refining workforce understand what is expected of them regarding process safety performance. (Baker et al. 2007: xii)

> ...

> BP's Board of Directors has been monitoring process safety performance of BP's operations based on information that BP's corporate management presented to it. A substantial gulf appears to have existed, however, between the actual performance of BP's process safety management systems and the company's perception of that performance. Although BP's executive and refining line management was responsible for ensuring the implementation of an integrated, comprehensive, and effective process safety management system, BP's Board has not ensured, as a best practice, that management did so. (Baker et al. 2007: xv)

One of the most important causal factors to emerge from the CSB Report and the Baker Panel review was the impact on corporate budget cuts on the safe operations at Texas City. The CSB observed that:

> Cost-cutting, failure to invest and production pressures from BP Group executive managers impaired process safety performance at Texas City. (CSB 2007: 25)

The impact of the budget cuts were significant, causing deficiencies in a range of areas that were important for effective safety management, areas such as training, maintenance, and staffing, yet the consequences of these cuts did not appear to have been considered before they were made, nor understood after they were made.

In civil legal proceeding in the United States, the then CEO of BP, Lord John-Browne, was examined by a lawyer, Brett Coon, acting on behalf of people who had lost family members in the disaster. The questioning goes directly to the issue of management's line of sight: what did senior management know about the impact that their decision making in the form of budget cuts have on safety at Texas City?

Mr Coon: Okay. Were you aware of any orders going to any of the refineries in the United States after the merger to cut their operational cost 25 percent?

Lord John-Browne: No, I wasn't, not in detail at all.

Mr Coon: Okay. Have you ever been provided with any of the lists of what the various individual refineries were doing in response to requests to reduce their budget after the merger?

Lord John-Browne: Not to my knowledge. I can't remember seeing such a list.

Mr Coon: Okay. Were you ever made aware of what BP Texas City did in response to the budget reduction requests out of London?

Lord John-Browne: Not—not to my knowledge.

Mr Coon: Were you aware that the plant manager at Texas City initiated dozens of different cost-cutting measures after receipt of this request?

Lord John-Browne: I wasn't aware of that.

Mr Coon: Were you ever made aware that they reduced the maintenance spend at BP Texas City as a result of that request?

Lord John-Browne: No, I wasn't aware of that.

Mr Coon: Were you aware that they reduced their staff operations in the control room at the ISOM unit by 50 percent as a result of that request?

Lord John-Browne: I wasn't aware of that.

Mr Coon: Were you aware that they curtailed, cut back or completely killed a number of training programs at BP Texas City as a result of that request?

Lord John-Browne: I wasn't aware of that. (John-Browne 2008: 38–39)

These comments and observations seem particularly pointed in light of the various findings into the organisational causes of the explosion, namely cost cutting and a lack of management oversight (CSB 2007: 25).

The Baker Panel Review, while not going to the precise level of Lord John-Browne's understanding of the impact of cost cutting on the organisation, was nevertheless able to say the following:

> The Panel recognizes that Browne is a very visible chief executive officer. Browne is generally noted for his leadership in various areas, including reducing carbon dioxide emissions and developing the use of alternative fuels. During the last eight years, Browne has spoken frequently on these issues across the globe. In 2005, The Financial Times named him the fifth most respected business leader in the world. Browne's passion and commitment for climate change is particularly apparent. In hindsight, the Panel believes that **if Browne had demonstrated comparable leadership on and commitment to process safety, that leadership and commitment would likely have resulted in a higher level of process safety performance in BP's U.S. refineries.** As observed in the 2003 Conference Board report on best practices in corporate safety and health, "[i]f the top executive believes in the worth of the strategies, sets expectations for other managers, follows through on those expectations, and commits appropriate resources, shared beliefs, norms, and practices will evolve." (Baker et al. 2007: 67) [emphasis added]

The nature of these questions and the responses raise a number of very interesting issues for managers to consider when looking at their individual responsibilities for safety and health. The first point to consider is the role of a CEO in events like BP Texas City, the level of knowledge that they should have, and even to some extent the fairness of the questions being asked.

I think that all three of these issues can be addressed in this way: Community standards seem clearly pointed towards an expectation that senior managers will have some sort of line of sight into the critical risks in their business. Whether or not Lord John-Browne understood the precise impact of the budget cuts on BP Texas City is probably not critical. However, it does seem that there is an expectation that the apparent failures within BP Texas City's safety management systems would have been bought to a sufficiently senior level of management, and eventually to the CEO if the matters had all remained unresolved, and they would then be dealt with. Again, this does not appear to be an expectation that senior managers are across the specific detail of every element of the safety management system, but it does seem clear that senior managers are expected to have assurance that critical health and safety risks in their business are controlled and that any failure in parts of the system designed to ensure that controls are effective to manage health and safety risks are bought to their attention.

This theme and expectation have certainly been reinforced in the recent disasters in North Western Australia and the Gulf of Mexico, the Montara and Deepwater Horizon events.

Montara

The Montara Commission of Inquiry found that the uncontrolled release of hydrocarbons on 21 August 2009 occurred because hydrocarbons were able to enter the casing string—the tubing that ran from the surface to the hydrocarbon reservoir under the sea bed, and through which the hydrocarbons would be extracted during normal production operations—and move to the surface without any barriers being in place.

The hydrocarbons were able to enter the casing string because of serious errors that had been made when cementing the casing shoe, a cement barrier at the bottom of the casing string that was designed to prevent hydrocarbons entering the casing string.

> Relevantly, the ... casing shoe had not been pressure tested in accordance with the company's Well Construction Standards, despite major problems having been experienced with the cementing job. In particular, the cement in the casing shoe was likely to have been compromised as it had been substantially over-displaced by fluid, resulting in what is known as a "wet shoe". (Borthwick 2010: 7)

Having entered the casing string, the hydrocarbons were able to escape because none of the other barriers designed to prevent just such an event were in place:

> Compounding the initial cementing problem was the fact that while two secondary well control barriers chosen by PTTEPAA—pressure containing anti-corrosion caps (PCCCs)—were programmed for

installation, only one was ever installed. Further, the PCCC that was installed (the 9⅝" PCCC) was not tested and verified in situ as required by the Well Construction Standards. (Borthwick 2010: 7)

...

When the West Atlas rig returned to the WHP in August 2009 it was discovered that the 13⅜" PCCC had never been installed. The absence of this PCCC had resulted in corrosion of the threads of the 13⅜" casing and this, in turn, led to the removal of the 9⅝" PCCC in order to clean the threads.

...

After the 9⅝" PCCC had been removed, the H1 Well was left in an unprotected state (and relying on an untested primary barrier) while the rig proceeded to complete other planned activities as part of batch drilling operations at the Montara WHP. The Blowout in the H1 Well occurred 15 hours later. (Borthwick 2010: 8)

Thankfully there was no loss of life, but hydrocarbons escaped unchecked into the environment until the well was finally "killed" on 3 November 2009, and the Montara Report was scathing in its criticism of the operations being conducted in the leadup to the accident:

The Inquiry has concluded that PTTEP Australasia (Ashmore Cartier) Pty Ltd (PTTEPAA) did not observe sensible oilfield practices at the Montara Oilfield. Major shortcomings in the company's procedures were widespread and systemic, directly leading to the Blowout. (Borthwick 2010: 6)

...

In essence, the way that PTTEPAA operated the Montara Oilfield did not come within a "bulls roar" of sensible oilfield practice. The Blowout was not a reflection of one unfortunate incident, or of bad luck. What happened with the H1 Well was an accident waiting to happen; the company's systems and processes were so deficient and its key personnel so lacking in basic competence, that the Blowout can

properly be said to have been an event waiting to
occur. (Borthwick 2010: 11)

A second report by the industry regulator in Australia, NOPSA,* has
been handed to the Commonwealth Director of Public Prosecutions to determine what, if any, charges should be laid for breaches of relevant legislation.†
There are any number of causal factors that are able to be drawn from
the Montara incident, and the Montara Inquiry was able to point to a
number of significant failures of PTTEPPA and other parties, including
the relevant regulator which contributed to the accident. Those failures
encompassed factors such as inadequate systems and procedures, the
failure to follow the procedures and standards that did exist, issues of
training and competence, cultures of production before safety, a failure
to identify and properly assess risk, and others, some of which will be
looked at in more detail in later chapters.

But from a management line of sight perspective, the questions that need
to be asked are these: What was happening at a senior management level
to identify those failures? Did management understand the risks associated
with suspending the wells and then preparing to commence production? Did
they understand the controls that were necessary to manage those risks, and
did they have any assurance that those controls were in place and effective?

Given what occurred and the findings of the Montara Inquiry, the
answers to all of those questions would seem to be no.

Andy Jacob, the Chief Operating Officer of PTTEPAA at the time of
the Montara incident, was cross examined by David Howe QC, Counsel
assisting the Montara Inquiry, on this aspect of management line of sight
and how senior management understood that work was being performed
in an acceptable manner.

Mr Howe QC: So whose responsibility was it in relation to ensuring that
the well construction department was actually administering its
affairs in accordance with PTT's expectations—was it the project
manager's responsibility?

Mr Jacob: At that time, yes.

Mr Howe QC: I want to suggest to you, sir, that if he shared your own view
about the issue of cost cutting, it is highly unlikely that he would have
brought any zealous attention to the daily drilling reports?

Mr Jacob: I'm sorry, can you explain the link?

Mr Howe QC: As I understand it, your evidence to the Commissioner was that
it was almost beyond your comprehension that cost cutting might occur
out on the rig in ways that might in any way compromise well control.

---

* The National Offshore Petroleum Safety Authority
† See http://www.nopsa.gov.au/

Mr Jacob: Yes.

Mr Howe QC: Do you recall that evidence?

Mr Jacob: Yes, yes, sorry.

Mr Howe QC: So if the project manager took a similar view, it is unlikely that he would have read these documents with any close attention, because he would have thought, "Oh, well, they will just do every single thing required of them"?

...

Mr Jacob: I wouldn't say that, necessarily.

Mr Howe QC: Do you agree, sir, that one of the things PTT needs to give very close attention to is a revisiting of this leisurely approach that senior management can rely on everyone to do everything exactly as required, because the very notion that there might be any cost cutting—or corner cutting, I'm sorry, is almost incomprehensible?

Mr Jacob: Yes and we identified that as an improvement in terms of the audits that we have suggested.

Mr Howe QC: And you will agree that that attitudinal approach about everyone being able to be relied upon not to cut corners is an attitude that arises from and infects organisations from the top down?

Mr Jacob: Yes, it would come from that, yes.

Mr Howe QC: Have you had any discussions with the current CEO of PTT to the effect, "Look, one of the risk factors we have identified is that at a corporate level, we were simply insufficiently vigilant, because we didn't do all that we could have and should have to ensure that there was no cutting of corners?"

Mr Jacob: He is aware of the recommendation; we have discussed the recommendation regarding the additional auditing, but I don't think I would have used the words "cutting corners".

Mr Howe QC: I'm not even talking about something like routine or random audits, sir. I'm just talking about the way the organisation goes about its day-to-day business that requires people who are given information to pay close attention to it; do you understand?

...

Mr Howe QC: And if the project manager had been sent a copy of the daily drilling report, presumably that wasn't an idle exercise and he should have paid it closer attention; do you agree?

Mr Jacob: Yes.

Mr Howe QC: How far up the chain do we have to go in order to establish the fact that there is a corporate problem of inattention so far as management of well control is concerned; do we need to go any further, sir?

Mr Jacob: You shouldn't do, no.

Mr Howe QC: So there is a widespread corporate cultural problem that involves reposing too much reliance upon those in the field and too little reliance upon a close consideration of information provided by them; do you agree?

Mr Jacob: I would rather say too much reliance on personnel below each of those people, be it offshore or onshore. I don't think it is restricted to offshore.

Mr Howe QC: Was the project manager a direct report to the CEO?

Mr Jacob: Yes.

Mr Howe QC: So, in all likelihood, we can go that one step further, too, can't we, sir, namely, that the CEO didn't properly inform himself of the nature and extent of the project manager's supervision of the affairs of the well construction department?

Mr Jacob: It would appear so, yes. (Jacob 2010: 1891–1893)

Unfortunately the CEO of PTTEPAA was not called as a witness in the proceedings, and so we get only a second-hand view of what measures were in place to provide him with assurance that critical safety risks were being managed at the time. But the themes contained in this passage of questioning are consistent with what is revealed in the various investigations into BP Texas City: Where is the evidence of management line of sight? How can senior management demonstrate that they have assurance that critical safety risks are being controlled?

One thing that does emerge very clearly from the cross examination in the Montara Inquiry is the reliance placed by management on personnel managing critical risks to do the right thing. The assumption that people will work safely.

If there is one thing that stands out from major accident investigations it is that managers are not entitled to assume that critical risks are being effectively managed. Positive steps must be taken and positive assurance obtained that those risks are being managed.

Not only is it the responsibility of organisations to understand that the critical health and safety risks are being controlled, but it is the expectation that all managers with responsibility for those critical risks will turn their minds to the management of those risks and be assured that they are being properly managed—all managers, up to and including the CEO, a point that was emphasised in investigations following the Deepwater Horizon event in the Gulf of Mexico.

Deepwater Horizon

On the evening of 20 April 2010, another, more serious uncontrolled release of hydrocarbons occurred, this time from the Macondo well, which formed part of a petroleum lease operated by BP Exploration and Production Inc. in the Gulf of Mexico. The hydrocarbons swamped Transocean's Deepwater Horizon drilling rig, and the resultant explosion and fire on the rig killed 11 people and injured 17 others. Hydrocarbons continued to flow from the well for some 87 days, resulting in, reputedly, one of the largest oil spills in US history.

Many of the safety management systems failures that were identified in the Montara Inquiry have been and continue to be identified in the Deepwater Horizon disaster, and again, the issue of management line of sight is being ventilated in various inquires.

On 14 June 2010, the United States Congress, House of Representatives, Subcommittee on Oversight and Investigations investigating the role of BP in the Deepwater Horizon explosion and oil spill wrote to the then CEO of BP, Tony Hayward, seeking information about a number of concerns that the committee had about practices they believed that BP had adopted that may have compromised safety (Waxman and Stupack 2010).

One of those concerns had to do with the practices that BP adopted when installing a final section of casing into the well:

> Well Design. On April 19, one day before the blowout, BP installed the final section of steel tubing in the well. BP had a choice of two primary options: it could lower a full string of "casing" from the top of the wellhead to the bottom of the well, or it could hang a "liner" from the lower end of the casing already in the well and install a "tieback" on top of the liner. The liner-tieback option would have taken extra time and was more expensive, but it would have been safer because it provided more barriers to the flow of gas up the annular space surrounding these steel tubes. A BP plan review prepared in mid-April recommended against the full string of casing because it would create "an open annulus to the wellhead" and make the seal assembly at the wellhead the "only barrier" to gas flow if the cement job failed. Despite this and other warnings, BP chose the more risky casing option, apparently because the liner option would have cost $7 to $10 million more and taken longer. (Waxman and Stupack 2010: 2)

On 17 June 2010, Mr Hayward gave testimony to the Committee, where there were very strong criticisms by some committee members of the role of senior management in the disaster. For example,

Mr Doyle: So we have reviewed all of their e-mails and communications. We find no record that they knew anything about this decision.* In fact, we find no evidence that they ever received briefings on the activities aboard the Deepwater Horizon before the explosion. These decisions all seem to have been delegated to much lower ranking officials.

Well, Mr Hayward, then, who was the one who made the decision to use a single tube of metal from the top of the well to the bottom? Who did make that decision?

Mr Hayward: I am not sure exactly who made the decision. It would have been a decision taken by the drilling organization in the Gulf of Mexico. They are the technical experts that have the technical knowledge and understanding to make decisions of that sort.

Mr Doyle: But you can't tell this committee who that person was?

Mr Hayward: I can't, sitting here today, I am afraid.

Mr Doyle: You can get this information to our committee? I mean, I think it is pretty amazing that this is the decision that had enormous consequences and you can't even tell the committee who made the decision on behalf of your company.

And the reason I am asking you these questions is because your industry is different than many. You are not the CEO of a department store chain where it is fine to leave decisions about running the store to branch managers. You know, if a department store middle manager makes a mistake, there are no life or death consequences.

What you do is different. You are drilling far below sea level into a region that is more like outer space than anything else. The consequences of that drilling are huge. If a mistake or misjudgement is made, workers on the rig can get killed and an environmental catastrophe can be unleashed.

The best minds in the senior leadership of a company should be paying close attention to those risks. *But it didn't happen here. And now we are all paying the consequences because* **those of you at the top don't seem to have a clue about what was going on on this rig.** (*Hayward 2010:* 129–130) [emphasis added]

Understanding management line of sight

One of the difficult aspects of understanding and managing line of sight is that there is seldom any direct causal link (in the sense that most people would understand that concept) between what a manager, well removed from the day to day operations of a business, knows about those operations and an accident.

---

* The decision to use a full string of "casing" from the top of the wellhead to the bottom of the well.

There is no clear causal link between the failures and risks alleged and the eventual outcomes and consequences that we eventually know occurred.

If a manager decides to remove a safety barrier and not replace it, it is relatively easy to understand why that manager's behaviour might be criticised. Intuitively, it is not so easy to understand why a chief executive officer should be held accountable for *that* failure.

The issue that managers need to turn their minds to is not so much the individual errors, but rather the cumulative failures and why they weren't recognised. Neither the Montara Incident nor the Deepwater Horizon incident was the result of a single, one off failure. Rather they were the result of longer term, systemic failures of safety management. The Montara Inquiry in particular was pointed in this regard—the admission that it was almost inconceivable to managers that corners might be cut when it came to safety goes directly to a failure of line of sight.

What are the critical health and safety risks in your business and how do you know that they are controlled?

It is not, in my view, the expectation that managers who are removed from day to day operations would have a close working knowledge of *everything* that is happening in their business at any given moment. It is, however, a reasonable expectation that they have some form of *positive* assurance that key safety controls are in place and effective, and this was not the case in either the Deepwater Horizon or Montara.

Although the precise combination of circumstances that arose in Montara and the Deepwater Horizon may have been unique to those factual scenario, the themes discussed, the nature of the questions asked of managers, and the implicit and explicit criticisms that arise are consistent across nearly all catastrophic events.

References

Baker, J. et al. 2007. *The report of the BP U.S. Refineries Independent Safety Review Panel.* U.S. Chemical Safety and Hazard Investigation Board, Washington. http://www.bp.com/liveassets/bp_internet/globalbp/globalbp_uk_english/reports_and_publications/presentations/STAGING/local_assets/pdf/Baker_panel_report.pdf.

Borthwick, D. 2010. *The report of the Montara Commission of Inquiry.* Montara Commission of Inquiry, Canberra. http://www.ret.gov.au/Department/Documents/MIR/Montara-Report.pdf. (accessed 25 November 2010).

CSB (U.S. Chemical Safety and Hazard Investigation Board). 2007. *Investigation report: Refinery explosion and fire.* Washington. http://www.csb.gov/assets/document/CSBFinalReportBP.pdf (accessed 23 November 2010).

Hayward, T. 2010. Transcript: U.S. House of Representatives, Subcommittee on Oversight and Investigations, Committee on Energy and Commerce. The role of BP in the Deepwater Horizon oil spill, 17 June 2010. Washington, D.C. http:// energycommerce.house.gov/documents/20100617/transcript.06.17.2010. oi.pdf (accessed 23 November 2010).

Jacob, A. 2010. *Transcript: Montara Commission of Inquiry.* http://www.montarain-quiry.gov.au/transcripts.html (accessed 29 September 2010).

John-Browne, Lord. 2008. Oral deposition dated 4 April 2008. http://www.texas-cityexplosion.com/etc/broadcast/files/08/BROWNE,%20JOHN-DEP.pdf (accessed 23 November 2010).

Waxman, A., and B. Stupak. 2010. Letter from the U.S. House of Representatives Subcommittee on Oversight and Investigations to Tony Hayward, Chief Executive Officer BP PLC dated 14 June 2010. http://energycommerce. house.gov/documents/20100614/Hayward.BP.2010.6.14.pdf (accessed 23 November 2010).

chapter 4

Understanding rules

> "Work rules and operating procedures are much less effective than they are normally believed to be."[*]

Introduction

Safety rules—processes, procedures standards and so forth—are one of the most ubiquitous features of modern safety management. They dominate the safety landscape. The suitability of safety rules and rule compliance are always a key and critical issue in a major accident inquiry. The development of more or new rules is a typical response to safety system failure.

Rules are seen as central to preventing accidents and achieving good safety outcomes.

In this chapter I will consider some of the difficulties associated with relying on rules to ensure workplace health and safety, how the reliance on rules has been examined in major accident investigations, and issues managers need to consider when relying on rules to both ensure a safe workplace and assure themselves that critical risks are being controlled.

Why rules don't always work

Notwithstanding the critical role played by rules, time and again they are found to be wanting for any number of reasons—for example because they are not capable of being understood, or because they simply do not exist for the types of risks that a worker may be faced with.

Over and above these fundamental problems, rules can and are routinely violated. Studies[†] have identified three main types of rule "violation":

- Routine violations—where shortcuts have crept into the way that work has been performed and have become the normal way of doing things, and in many cases are not even recognised as being in breach of the rules.

---

[*] Hollnagel, Woods and Leveson 2008: 327
[†] See for example Reason, Parker and Lawton 1998: 289–304

- Situational violations—where the work cannot be done if the rule has to be complied with. The requirement to use a tool lanyard on the torque wrench as discussed in Chapter 1 is such an example; it was a rule that could not be complied with because a suitable lanyard did not exist.
- Exceptional violations—where the immediacy of a situation provokes a response, sometimes in breach of a life-saving rule, such as a worker entering into a confined space to rescue a collapsed colleague without applying any safe-entry protocols, or a mine manager who rushes into a mine following a rock fall because of concern for missing workers.

Professor Andrew Hopkins (2005) gives an interesting overview of and insight into a rules-based safety culture in the context of the Glenbrook train crash, which occurred in New South Wales, Australia, in December 1999.

Management of rail safety in Australia, like in many other counties, is heavily reliant on rules as a primary means of preventing accidents, but as Hopkins describes, there are many limitations associated with a reliance on rules—not least of which is that people might not follow them. Some of the limitations of the use of rules identified by Hopkins and others include:

- A lack of clarity around rules, meaning that they may be open to misinterpretation, or even to being ignored if they are too complex.
- The tendency for rules to have an impact on performing work efficiently, meaning that workers will develop "work arounds" in order to avoid the rules in the name of getting the job done.
- The risk of heavily regulated workplaces creating a lack of risk awareness as workers come to rely on rules as the measure of safety. On the one hand, a belief that "If I follow the rules I must be safe", and on the other, "If there are no rules it must be safe because otherwise there would be a rule."
- The use of rules to assign blame, rather than manage risks.

In some cases a strong reliance on rules can provide an effective mechanism for ensuring safety in the workplace, but this requires strict adherence to the rules, often at the expense of efficiency and flexibility and with the risk that any deviation from the rules can result in accidents. This dichotomy of rule reliance and rule deviation can co-exist within an organisation, as can been seen from a study of European rail safety:

> Passenger safety appears to be achieved by defining very clearly in advance what are the necessary prerequisites of safe operation and forbidding

> operation outside them. ... Then the system moves
> outside this clearly defined safe envelope, the rail-
> way system stops, regroups and restarts only when
> the necessary operating conditions have been re-
> established. Hence safety is achieved by sacrificing
> goals, traffic volume and punctuality. The system
> does not achieve all of its goals simultaneously
> and flexibly and is not resilient. Yet it achieves a
> very high safety performance on this performance
> measure.
>
> On track maintenance worker safety, the system
> has a poorer performance. ... The trade-off works
> here in the other direction. Punctuality of the trains
> and traffic volume are bought at the price of safety.
> The system is not resilient and performs relatively
> poorly. (Hale 2008: 146–147)

A possible indicator of the effectiveness or otherwise of safety management is the gap between work as management imagines it and work
as actually performed (Hale 2008), and one of the risks inherent in safety
rules is that they can contribute to this gap—what Borys described as the
illusion of safety (Borys 2009). Dekker (2008: 86) illustrates the problem
as follows:

> Commercial aircraft line maintenance is emblem-
> atic: A job-perception gap exists where supervi-
> sors are convinced that safety and success result
> from mechanics following procedures—**a sign-off
> means that applicable procedures were followed**.
> But mechanics may encounter problems for which
> the right tools or parts are not at hand; the aircraft
> may be parked far away from base. Or there may be
> too little time: Aircraft with a considerable number
> of problems may have to be turned around for the
> next flight within half an hour. Mechanics, conse-
> quently, see success as the result of their evolved
> skills at adapting, inventing, compromising, and
> improvising in the face of local pressures and chal-
> lenges on the line—**a sign-off means the job was
> accomplished in spite of resource limitations,
> organisational dilemmas, and pressures**. [empha-
> sis added]

BP Texas City

A more dramatic example arose in the BP Texas City refinery, where one of the key contributing factors in the disaster was the failure of various alarms and indicators to warn of impending danger:

> A functionality check of all alarms and instruments was also required prior to startup, but these checks were not completed. On March 22, 2005,* instrument technicians had begun checking the critical alarms when a supervisor told them that the unit was starting up and there was no time for additional checks.
>
> While some alarms were tested, most were not prior to startup. The supervisor, however, initialled on the startup procedure that those checks had been completed. (CSB 2007: 48–49)

Montara

The gap between what was imagined and what was actually happening was well illustrated in the course of the Montara inquiry when PTTEPAA's Chief Operating Officer gave evidence that it was beyond his comprehension that shortcuts might occur out on the rig that could in any way compromise well control.

Mr Howe QC: No, I mean would he also have taken the position that he couldn't, as it were, credit that corners might be cut or **people might lose sight of proper procedures** because they were diverted to endeavours to save time and money, and the like?

Mr Jacob: I would think so, but obviously I can't talk for him.

Mr Howe QC: What about your CEO—do you think the CEO shared that same approach?

Mr Jacob: Again, I don't think anybody in the organisation would credit that things would be done to the detriment of safety for the benefit of cost.

Mr Howe QC: I want to suggest to you, sir, that that very evidence reveals a problem, namely, that no-one in the organisation seems to have properly credited the known phenomenon that when people are pursuing efficiencies and time savings and cost savings, **they can lose sight of the need to observe proper procedures**.

Mr Jacob: Sorry, could you repeat the first part of that?

* The day before the accident

Mr Howe QC: Yes. You seem to be saying that, to your knowledge or understanding, no-one in PTT would have credited at the time that people involved in well management and well control might have succumbed to any sort of corner-cutting or inattention to proper procedures by virtue of the desire to achieve time and cost savings.

Mr Jacob: Mmm-hmm, yes.

Mr Howe QC: I'm suggesting to you that the very fact that you are giving that evidence identifies a problem, namely, senior management did not properly recognise the plain fact of ordinary human nature and a known phenomenon, namely, when you have lots of people applying themselves to achieving time and financial efficiencies, **they can lose sight of the need to properly attend to processes**.

Mr Jacob: On the basis that there weren't systems in place to ensure that the barriers, et cetera, were identified as being in place and verified and that, yes, I can accept that. (Jacob 2010: 1784) [emphasis added]

What this examination highlights is the need for safety management to take account of the fact that people will, for whatever reason, not always comply with safety rules. Systems, and managers, need processes to be able to identify when rules are not being followed. These processes might include audits, investigations, management walk-arounds, safety conversations and so on. When rule violations are identified, managers need to resist the urge to see the violation as an aberration, as a unique, one-off occurrence or something linked to the particular individual breaching the rule at the time. The response to broken rules needs to be to dig deeper—to be assured that the specific rule violation is not symptomatic of deeper, more fundamental underlying weaknesses in the safety management of the organisation.

Texting and driving

In some cases, the rule itself can be the problem, having unintended consequences that increase the very risks that it was designed to manage. A study of jurisdictions in the United States that have banned texting while driving as a safety measure suggests that the ban may have actually *increased* risks to road users, rather than diminished them:

> This unexpected consequence of banning texting suggests that texting drivers have responded to the law, perhaps by attempting to avoid fines by hiding their phones from view. If this causes them to take their eyes off the road more than before the ban, then the bans may make texting more danger-

ous rather than eliminating it. (Highway Loss Data
Institute 2010: 8)

Yet reliance on rules is still arguably the pre-eminent philosophy
underpinning many safety management systems, and "operator error"
is the most basic manifestation of rule reliance as a safety management
philosophy—the accident occurred because the worker did not follow the
rules.

A manager's obligation is not only to ensure that appropriate rules
are in place to manage the risks they are responsible for, but also to know
that those rules are effective in managing the risks; and rules may not be
effective for any number of reasons, not least of which because they are
not followed.

The Herald of Free Enterprise

It is important that rules, as far as possible, contribute to effective safety
management and not undermine it. However, some rules are just bad
rules and have the potential to actively contribute to accidents.

On 6 March 1987, the roll on/roll off ferry *Herald of Free Enterprise* was
sailing from Zeebrugge, Belgium, bound for Dover in the UK when it capsized and partially sank, resulting in the deaths of 150 passengers and 38
crew. The Secretary of State for Transport ordered a formal investigation
into the tragedy, and the formal inquiry commenced on 27 April 1987, only
seven weeks after the incident. The findings of the inquiry were issued on
24 July 1987 (Department of Transport 1987).

The inquiry found that the *Herald of Free Enterprise* capsized because
she had gone to sea with the inner and outer bow doors open. Although
there were a raft of organisational and management failures that contributed to the disaster, which will be examined in later chapters, the role of
rules was also considered in the inquiry.

The rules governing the operation of the fleet identified in the *Herald
of Free Enterprise* inquiry relied primarily on Ship's Standing Orders. One
such order was order 01.09: 01.09

> Ready for Sea Heads of Departments are to report to
> the Master immediately they are aware of any defi-
> ciency which is likely to cause their departments to
> be unready for sea in any respect at the due sailing
> time.
>
> **In the absence of any such report the Master
> will assume**, at the due sailing time, that the

> vessel is ready for sea in all respects. (Department
> of Transport 1987: 12) [Emphasis added]

Even leaving aside the fact that the critical technical failure causing the disaster was to sail with the inner and outer bow doors open, the risks inherent in any assumptions about readiness to sail ought to have been apparent.

Whether or not this rule was actually at the front of the Master's mind at the time he made the decision to sail away from Zeebrugge was not explored in the inquiry, but at best it is indicative of an organisation that has not properly considered the risks inherent in its operations. At worst, it is a rule that positively undermines prudent safety management.

We shall see further on that assumption typically plays a critical role in most major accidents, and it is bad enough when managers make assumptions about the management and effectiveness of safety, much less when the rules of an organisation *institutionalise* assumptions about safety.

In isolation, it is difficult to understand why a rule could be crafted to allow such an important aspect of operations, being ready to go to sea, to be left to assumption rather than positive assurance. It may be that the rule had become subject to misinterpretation over time, with the inquiry finding:

> That order was unsatisfactory in many respects. It followed immediately after 01.08 which was an order that defects had to be reported to the Head of Department. The sequence of orders raises at least a suspicion that the draftsman used the word "deficiency" in 01.09 as synonymous with "defect" in 01.08. On one construction of the orders, order 01.09 was merely completing the process of ensuring that the Master was apprised of all defects. That is how this Court would have interpreted it. But it appears that that is not the way in which order 01.09 was interpreted by deck officers. Masters came to rely upon the absence of any report at the time of sailing as satisfying them that their ship was ready for sea in all respects. That was, of course, a very dangerous assumption. (Department of Transport 1987: 12)

If this was the case, it is just as likely that the misinterpretation was a result of an organisation that did not prioritise safety.

Western Power*

Western Power was an electricity power utility providing electricity to Western Australia, including remote and regional areas of the State.

In January 2000 two employees of Western Power, Mr Pike, an experienced linesman, and Mr Tan, a graduate engineer, entered a remote electricity substation. At the time, Mr Pike was carrying out his regular duties for Western Power and Mr Tan had come along with him to gain experience about Western Power's operations. At the substation Mr Pike had planned to read a meter at the facility when the two men noticed a sapling that was growing near one of the transformers in the substation. Undergrowth in the substations can cause problems, and it needs to be cleared from time to time.

Although it was not a part of Mr Pike's job, he and Mr Tan decided to clear the undergrowth, believing if they did not, it would be some time before Western Power personnel could return to do the work. They were acting in what they believed to be the best interests of Western Power, as the Court found:

> They decided to clear the sapling there and then and
> so save someone else the trouble of doing it later.†

But while in the substation, Mr Tan came into contact with a live transformer, was electrocuted, and died.

The substation was inside a pumping station compound operated by another State utility, and both the pumping station and the substation were fenced and only accessible through locked gates. Only Western Power personnel with an appropriate key could gain access to the pumping station, and only Western Power Corporation personnel with a particular key, known as an NK6 key, could gain access to the electricity substation. Under Western Power's work procedures, only authorized Western Power personnel were permitted to enter the electricity substation.

Mr Pike had the key that allowed him to enter the substation.

The Western Power case provides an interesting contrast to a number of cases involving employees who deviate from the rules, insofar as Mr Pike fully understood the requirements and the expectations placed on him. This was not a case of a worker who misunderstood the rules, as Mr Pike was fully aware that he was not authorized to enter the substation:

---

* Western Power Corporation v Shepherd (2004) WASCA 233 (Western Power)
† Ibid, [46]

> The Magistrate found that neither Mr Pike nor
> Mr Tan contacted anyone in Western Power about
> whether they should enter the substation prior to
> doing so. He also found that Mr Pike was aware that
> it was necessary to obtain a permit to enter the sub-
> station. The Magistrate accepted that it was contrary
> to all instructions given to Mr Pike, and the training
> he had received, to enter the substation enclosure.
> He noted that Mr Pike could not explain why he did
> so and had stated he had never done so before.*

On the face it, Western Power had done all that they could be expected
to do to manage the risks associated with personnel entering the substation, having regard to the systems employed and the permit system in
place.

From another perspective, the systems used by Western Power:

> ... left room for mistakes, error and misjudgement
> by a wide range of employees with different range
> of capacities and experience ...†

The limit to which rules can be relied on to manage safety is particularly apparent when there are other, more effective controls that could
have been implemented.

> Notwithstanding Western Power Corporation's sen-
> sible procedures and systems, ... on the evidence
> before the Magistrate, it was open to him to find that
> it was conceivable that an experienced linesman,
> such as Mr Pike, in the course of carrying out [his]
> duties ... might decide ... to enter the electricity
> substation for the purpose of removing some veg-
> etation that might eventually prove problematic to
> the operation of the substation and so save some
> other Western Power employee the trouble of trav-
> elling to Coolgardie to perform what appeared to
> be a relatively simple task ...
>
> ...
>
> There was no doubt, on the evidence before
> the Magistrate, that the means were available to

* Ibid, [17]
† Ibid, [18]

> Western Power, and were well within their knowl-
> edge and financial reach, to avoid any hazard to an
> employee who might enter the electricity substa-
> tion ...*

What the Western Power case demonstrates, and reminds managers of, is that rules cannot be relied on to manage safety in the face of inherent risks in the workplace that can be reduced by other means, such as engineering controls. Safety management systems need to recognise that people make mistakes—often well-intentioned mistakes made with the benefit of the employer or fellow workmates in mind, but mistakes nonetheless—and those systems need to be capable of accommodating, as far as possible, the errors that people might make. After all:

> The duty to provide a risk free work environment is
> a duty owed not only to the careful and observant
> employee but also to the hasty, careless, inadvertent,
> inattentive, unreasonable or disobedient employee
> in respect of conduct that is reasonably foreseeable.†

Black Hawk 221

On 29 November 2006, Australian Army Black Hawk helicopter A25-221 (Black Hawk 221) was participating in a training flight as part of operations in the southern Pacific Ocean when it crashed into the deck of *HMAS Kanimbla* and fell into the sea, resulting in the deaths of two personnel, the pilot, and one of the passengers.

The subsequent Defence Force Board of Inquiry (Levine, Fielder and Rourke 2008) found that the cause of the crash was pilot error by the aircraft captain, but that the finding:

> ... cannot be viewed in isolation nor to be taken as
> the attribution of specific and discrete blame against
> one serving officer of considerable experience and
> unquestioned reputation (Levine et al. 2008: xix).

In looking beyond operator error the Board of Inquiry found:

> Thus the tragic events of 29 November 2006 can be
> viewed and are viewed by this Board as the culmi-
> nation of an uncontrolled, inadequately supervised

---

* Ibid, [47–49]
† Workcover Authority of NSW v TRW (2001) NSWIRComm 52 [at 13] (TRW)

> and indeed unnoticed evolution of **normalized deviance** coupled with an unchecked level of complacency. (Levine et al. 2008: xxi) [emphasis added]

Normalized deviance is a working example of what I discussed at the start of the chapter—*routine violations* where shortcuts have crept into the way that work has been performed and have become the normal way of doing things. It was described to the Board of Inquiry as:

> Over time, operational personnel develop informal and spontaneous group practices and shortcuts to circumvent deficiencies in equipment design, clumsy procedures or policies that are incompatible with the realities of daily operations, all of which complicate operational tasks. These informal practices are the product of the collective know-how and hands-on experience of a group, and had they eventually become normal practices. This does not, however, negate the fact that they are deviations from procedures that are established and sanctioned by the organisation, hence that term "normalization of deviance". In most cases normalized deviance is effective, at least temporarily. However, it runs counter to the practices upon which the system operation is predicated. In this sense, like any shortcut to standard procedures, normalized deviance carries the potential for unanticipated "downsides" that might unexpectedly trigger unsafe situations. (Levine et al. 2008: 101)

The Board of Inquiry examined a number of examples of what it considered to be normalized deviance, one of which was an acceptance of operating with a tailwind as "normal". Relevant standards helicopter operation—the rules—stated that aircrew must always *"operate into the wind unless a manoeuvre calls for an out of wind condition"* (Levine, Fielder and Rourke 2008: 105). However, the evidence demonstrated what the Board of Inquiry described as *"nonchalance with respect to wind"* (Levine et al. 2008: 102):

Q: Were you aware at the time and that you were doing the walk-through that if you came down the active side port side and terminated to a hover over the flight deck, you will be terminating out of wind?

A: No, I was not.

Q: When you conducted your walk-through, did anyone raise with you that as had terminated turning left over the flight deck, something along the lines of, "Gentlemen, we will now have a tailwind"?

A: No Sir.

Q: Is that something that would have been of importance to you?

A: Myself, sir, I don't believe it's of great importance, no.

Q: Why is that?

A: Because we can still adjust the approach as required to counter for any tailwind or wind effect on the aircraft on the day. (Levine et al. 2008: 103)

The Black Hawk 221 Board of Inquiry is particularly instructive to managers who place a high reliance on rules to manage risks in their business and ensure safety. If such a highly regulated, rule-orientated, disciplined and professional function as Army Aviation can drift into a state of noncompliance with rules and resulting tragedy, what is it about your organisation or workforce that makes it immune from similar error? What assurance do you have that the rules in your business are actually operating as intended and are effective in managing the risks that they are designed to manage?

As the Board of Inquiry observed, and as the history of major accidents reinforces, instances of normalisation are not uncommon:

> The phenomena of normalized deviance is not the exclusive problem and concern of just 171 [aviation squadron]. It is common in all walks of life and certainly there would be many other areas of the ADF were examples could be found. The issue becomes how to avoid it and this comes down to leadership, adherence to regulations and rolls, affective audits and checks from external agencies, a healthy safety culture and vitality; the effective application of operational risk management. (Levine et al. 2008: 130)

Final thoughts about rules

Rules are necessary. But rules are just one tool that organisations use to help manage the health and safety risks of their business. They are not a cure-all.

In order to be effective it must be recognised that rules have limits. There is a level or irreducible uncertainty in life; while we must make every effort to put in place systems that are as effective as possible, we will never foresee and predict all risks or imagine every permutation of human behaviour that could impact on safety in the workplace.

Managers must be rigorous and vigilant in challenging the rules in their organisation, constantly questioning whether the rules are as effective (and understood) as they can be in minimising risk. At the same time, because of the inherent deficiencies in rules and in people, including ourselves, it is essential always to ask whether our workplaces and processes are designed to minimise the need for rules—are they error tolerant so that if rules are not in place or mistakes do get made, those gaps and mistakes do not lead to tragedy. And finally, do we equip our people with the skills, provide them with the supervision, and create a culture that will allow them to recognise and react to potential warning signs in their work so that even in the absence of rules, or a proper understanding of the rules, they have an ability to recognise, or at least question the possibility of, potential danger emerging and pull back from the brink?

References

Borys, D. 2009. Exploring risk-awareness as a cultural approach to safety: Exposing the gap between work as imagined and work as actually performed. *Safety Science Monitor*, Issue 2, Volume 13. http://ssmon.chb.kth.se/vol13/issue2/3_Borys.pdf (accessed 23 November 2010).

CSB (U.S. Chemical Safety and Hazard Investigation Board). 2007. *Investigation report: Refinery explosion and fire*. Washington. http://www.csb.gov/assets/document/CSBFinalReportBP.pdf (accessed 23 November 2010).

Dekker, S. 2008. Resilience engineering: Chronicling the emergence of confused consensus. In Hollnagel, E., D. Woods and N. Leveson. *Resilience engineering concepts and precepts*. England: Ashgate.

Department of Transport. 1987. *MV Herald of Free Enterprise report of Court No. 8074 Formal Investigation*. London. http://www.maib.gov.uk/publications/investigation_reports/herald_of_free_enterprise/herald_of_free_enterprise_report.cfm (accessed 23 November 2010).

Hale, A., and T. Heijer. 2008. Is resilience really necessary? The case of railways. In *Texting laws and collision claim frequencies*, Vol. 27, No. 11 September 2010 Highway Loss Data Institute Bulletin. http://www.iihs.org/research/topics/pdf/HLDI_Bulletin_27_11.pdf (accessed 23 November 2010).

Hollnagel, E., D. Woods and N. Leveson. 2008. *Resilience engineering concepts and precepts*. England: Ashgate.

Hopkins, A. 2005. *Safety, culture and risk: The organisational causes of disasters*. Australia: CCH.

Jacob, A. 2010. *Transcript: Montara Commission of Inquiry*. http://www.montarainquiry.gov.au/transcripts.html (accessed 29 September 2010).

Levine, D., S. Fielder and A. Rourke. 2008. *Report of the Board of Inquiry into the deaths of Captain Mark Bingley and Trooper Joshua Porter following the loss of Army Black Hawk 221 on 29 November 2006 whilst deployed in HMAS KANIMLA for Operation QUICKSTEP*. http://www.defence.gov.au/coi/reports/bh221_boi_report.pdf (accessed 21 November 2010).

Reason J. T., D. Parker and R. Lawton. 1998. Organisational controls and safety: The varieties of rule-related behaviour. *Journal of Occupational and Organisational Psychology.* 71: 289–304.

chapter 5

Training and competence

"Where an accident occurs, one of the first questions for the employer should be whether the event was the result of the employer's failure to provide the necessary instruction, training, supervision and monitoring of its employees, to ensure compliance with its safe system of work."[*]

Introduction

There can be little doubt, nor would I expect that there would be any meaningful disagreement, that training and competency are significant elements of safety management and that the level of training and competence within an organisation goes a long way to determining the quality and effectiveness of that organisation's safety management system. In many jurisdictions the need to ensure adequate training and competence is embodied in various legislation, regulations, or standards. For example, the Western Australian Mines Safety and Inspection Act of 1994 provides:

> An employer must, so far as is practicable, provide and maintain at a mine a working environment in which that employer's employees are not exposed to hazards and, in particular, but without limiting the generality of that general obligation, an employer must—
>
> ...
>
> (b) provide such information, instructions and training to and supervision of employees as is necessary to enable them to perform their work in such a manner that they are not exposed to hazards; ...[†]

In this chapter I will consider the role of training and competence in preventing major accidents, and how an organisation's training and

[*] R v Commercial Industrial Construction Group Pty Ltd (2006) VSCA 181, [45] (CICG)
[†] Section 9

63

competence programs are examined following a major accident. I will also consider the limitations of relying on training and competence to ensure a safe workplace.

Relying on training and competence

There is a very close and obvious relationship between training and competence on the one hand, and "rules" on the other, in the context discussed in Chapter 4. On one level, the relationship is very direct; people need to be trained on the rules in order to be able to understand, implement, or follow them. On another level it is important that managers understand that extensive training and competency programs, of themselves, do not provide assurance that the critical risks in a business are being effectively managed.

To be effective, training must be:

1. Provided
2. Understood
3. Effectively implemented
4. Enforced

In some cases adequate training can be relied on to show that an organisation has met their *legal* obligations.

In the Prospect Electricity case* a linesman and a lineworker who commenced work on high voltage power lines without an access permit verifying that the job was isolated were electrocuted, one of them fatally. In proceedings arising from the fatality the prosecution alleged that the employer did not provide adequate supervision. In response, the Court said:

> It is a version of the infinite regression of supervision argument, which if accepted would mean that every well trained tradesman should be supervised or observed by a superior trained tradesman, and he seemingly in turn should be supervised himself by someone even higher.
>
> The "lack of instruction" submission relied upon founders on the same practical argument. No fully trained and experienced tradesman needs to be told to cross the road safely, or to be accompanied by a supervisor to ensure that he does so.

* Inspector Davies v Prospect Electricity (unreported, Fisher P, CT747, 9 Nov 1992) (Prospect Electricity)

> Similarly, a thoroughly trained and qualified tradesman, employed **after extensive instruction, the adequacy of which again is not criticised**, does not need to be advised not to grasp an unprotected wire energised by 11KCV, nor in the real world, should supervisors be on standby in case he does.
>
> ...
>
> ... that the sole cause of the accident was that through some **inexplicable inadvertence** the two men did not attend to a safe system of work which they had undoubtedly been thoroughly taught and for many years had consistently followed, which failure led directly to a fatal accident which was in the full sense of the word, an accident. No other explanation would appear to be available.[*] [emphasis added]

There are of course two difficulties with this view. First, even if you have met your legal obligations, an accident has happened, people have died, and the personal and business impact of that has to be managed—clearly if preventing that outcome means exceeding the minimum legal expectations, then there are benefits that go far beyond simple legal compliance.

Second, how can you be sure that you have met the expectations that attach to training and competency? How do you know that training has been provided, understood, and is effective and enforced?

As with most of the topics discussed in this book, training and competence must be directed at the risks in the business—in particular the critical risks—and often this is not the case. Often training is directed to what is "known" or "simple", and effective training to manage abnormal situations (often the critical risk) is not provided.

Moreover, what the cases continue to reveal is that, time and again, training is not adequate for any one of (or more typically a combination of) the factors listed above.

Longford

Training and competence of operators was a significant issue in dispute during the inquiry into the ESSO Longford gas plant explosion, with ESSO taking the view that operators had information available to them to manage the risks:

[*] Ibid

> ESSO challenged the evidence of the operators … It
> relied upon OIMS, its operator training programs,
> the Red Book and expert evidence from witnesses
> … that the loss of lean oil circulation was a funda-
> mental issue to be addressed in GP1. He said that
> operators failed to do things on the day of the acci-
> dent "that are so basic in a lean oil plant … and so
> standard in the industry and in a plant of that type
> …". He said that hazards associated with the loss of
> lean oil flow were well known. (Dawson and Brooks
> 1999: 197)

The findings of the Royal Commission, however, differed markedly
from the position taken by ESSO, concluding that the "real cause" of the
accident was training:

> Notwithstanding the matters mentioned above,
> the conclusion is inevitable that the accident
> which occurred on 25 September 1998 would not
> have occurred had appropriate steps been taken
> following the tripping of the GP1201 pumps. …
> Those who were operating GP1 on 25 September
> 1998 did not have knowledge of the dangers asso-
> ciated with loss of lean oil flow and did not take
> the steps necessary to avert those dangers. Nor
> did those charged with the supervision of the
> operations have the necessary knowledge and
> the steps taken by them were inappropriate. The
> lack of knowledge on the part of both operators
> and supervisors was directly attributable to a
> deficiency in their initial or subsequent training.
> Not only was their training inadequate, but there
> were no current operating procedures to guide
> them in dealing with the problem which they
> encountered on 25 September 1998. (Dawson and
> Brooks 1999: 234)

The Royal Commission found that the collective experience of those
present at the gas plant on 25 September 1998 was more than 200 years at
Longford, but no one recognised the hazards associated with the plant
leading to the explosion and fire. Examples of this lack of experience and
knowledge included Jim Ward, the GP1 Control Room Operator on the
day of the incident, who had been employed by ESSO for 18 years with
11 years experience as an operator. The Royal Commission found that

his evidence presented a less than adequate understanding of the processes involved and in particular of the consequence of process upsets.* Mr Ward's evidence, consistent with that of many other witnesses, was that he did not appreciate the safety consequences of the loss of lean oil flow and that his understanding was that if the lean oil flow system did not operate, the plant would produce off-specification product.

Another of the consequences of the explosion was that ESSO Australia was prosecuted under Victorian Occupational Safety and Health legislation, and, unsurprisingly, the issue of training and operator knowledge was again the subject of significant focus.

> How unfortunate it was that, on Esso's instructions, its solicitors submitted to the Royal Commission:
>
> *"Mr Ward was in possession of the necessary information to initiate appropriate action to address the loss of lean oil circulation ... Mr Ward's failure in this respect was due to reasons peculiar to himself."*
>
> (Submissions on behalf of Esso Australia by its solicitors to the Longford Royal Commission, 26 April 1999, p. 83, paragraph 269.) The truth is there was only one entity responsible for lack of knowledge on that day: Esso. It, and it alone, should have properly trained the operators and supervisors not only in production, which it did, but also in safety. It, and it alone, failed to do so. Mr Ward and the employees did not fail. Esso failed.†
>
> ...
>
> It was evident from the evidence given by witness after witness before me that the loyal employees, including supervisors, of Esso were entirely unaware of the deadly danger lurking at GP 905 on the Friday morning, 25 September 1998, particularly around 12 noon. They were loyally attending to a leak in GP 922 and evident cold on GP 905 and related areas. Only one man knew the dangers. Mr Vandersteen, a fitter in the maintenance section, saw what was evident to be seen and in evidence said this: "I just said, "Fuck this, I'm out of here."

* See, generally, Dawson and Brooks 1999: 195–198.
† Director of Public Prosecutions v Esso Australia Pty Ltd (2001) VSC 263 [44] (Esso Australia)

> We jumped on our bikes and we left the area."
> The explosion occurred immediately thereafter.
> Mr Vandersteen was not trained by Esso, but was
> trained by the Navy. It was the Navy, not Esso, who
> taught him to be aware of such danger. This failure
> to train in safety is a most serious dereliction.*

In making recommendations, the Longford Royal Commission noted:

> Of central importance is the training by Esso of its
> employees. An obligation should be imposed upon
> Esso to demonstrate that its training programs and
> techniques impart a knowledge of all identifiable
> hazards and the procedures required to deal with
> them. Not only should Esso be required to demon-
> strate that the necessary knowledge is imparted,
> but also that it is retained for use in an emergency.
> (Dawson and Brooks 1999: 240)

As we shall see later in this chapter, the failure to train workers about identifiable hazards and ensuring that their knowledge is retained for use in an emergency is a common theme in many major accident events.

BP Texas City

We have already described some of the elements of the Texas City refinery explosion in March 2005, which was one of the worst industrial accidents in US history. While there was a complex combination of causal factors that were ultimately found to be at play, one of the most critical "facts" leading to the explosion was operations personnel overfilling a raffinate splitter tower with hydrocarbons during start-up procedures. The hydro-carbons eventually overflowed from the tower and released so that they were able to ignite.

The hazard of overfilling the tower

The risk of overfilling the tower was a critical risk, yet training in scenarios that might lead to the tower overfilling or other abnormal situations was not adequately covered. Indeed, the start-up procedures themselves and the start-up process are recognised as being more hazardous operations than "steady state" operations, but again the training about the potential hazards was found to be inadequate.

---

* Ibid [12]

Inadequate training for operations personnel, particularly for the board operator position, contributed to causing the incident. The hazards of unit startup, including tower overfill scenarios, were not adequately covered in operator training.

BP Board Operator Training consisted of new-hire basic operator training; a two-day generic troubleshooting course; computer-based tutorials; and on-the-job training.

The ISOM unit operator training program did not include

- training for abnormal situation management, the importance of material balance calculations, and how to avoid high liquid level in towers;
- effective verification methods of operator knowledge and qualifications; and
- a formal program for operations crews to discuss potentially hazardous conditions, such as startup or shutdown, to enhance operator knowledge and define roles. (CSB 2007: 94)

Indeed the hazard of overfilling the tower seemed to be so far from the minds of operators that they had actually developed a practice of *deliberately* overfilling the tower. Operators were concerned that if the level of hydrocarbons inside the tower dropped too low during startup, it might cause damage to equipment (CSB 2007: 73–76).

They were not being deliberately unsafe, but a combination of poor rules and normalisation led to the development of unrecognised unsafe practices.

This comment from Texas City very closely reflects the comment of the Longford Royal Commission when considering the critical hazard in the Longford explosion—the loss of warm lean oil:

At no relevant time did any program include training with respect to the hazards associated with the loss of lean oil flow, the hazards associated with the uncontrolled flow of condensate into the rich oil stream from the absorbers, the critical operating temperatures … , the circumstances in which brittle fracture might occur or the procedures for the shutdown or start up of [Gas Plant 1]. (Dawson and Brooks 1999: 193)

Computer-based training

As part of the response to budget cuts within Texas City, refinery management took a number of steps that impacted directly on training, and by 2003 an external audit identified that training was not *"up to the challenge of performance expectations and anticipated turnover"* (CSB 2007: 99).

One of the areas impacted by the cost cutting was a move towards computer-based training:

> Corporate-level decisions, such as budget and staff reductions, impaired the delivery of training at the Texas City site. Between 1998 and 2004, the budget for the Texas City refinery L&D department was cut in half, from $2.8 to $1.4 million, and its staff was reduced from 28 to eight. At the time of the incident, four PSM positions, which could have assisted with training and gun drills, were also vacant in the West Plant of the refinery (which included the ISOM Unit).
>
> To make up for fewer L&D trainers, BP Texas City went to computer-based training for policies, procedures, and process unit lessons. This type of training saved the company money, according to the individual who, at the time of the incident, was head of the L&D: *"[computer-based training] was definitely a cost decision … made across the site; we will push computer based training to you as opposed to bringing you the classroom training … it was a business decision to minimize costs."* However, operators who require training for abnormal conditions would not benefit from computer-based training that often focuses on memorizing facts, not troubleshooting unusual events. (CSB 2007: 98)

The Baker Panel review formed the view that computer-based training contributed to inadequate process safety training. They noted:

> Computer-based training, while useful for informing personnel about changes and effective in teaching certain types of subject matter, is less effective at developing adequate process safety awareness and the skills and ability needed to apply knowledge in actual operations. (Baker et al. 2007: 164)

The Baker Panel identified that computer-based training was not rigorous enough and did not adequately assess a worker's overall knowledge and skill level. It was described by refinery personnel as easy to pass, and passing did not always accurately represent and employee's abilities in the field.

> For example, a 2001 root cause analysis report relating to an incident at the Toledo refinery involving an operator stated that the training system does not fully assess if a person has mastered the material being taught, noting that the operator in question passed a training test with a 100 percent score.
>
> Although employees must eventually pass the computer-based tests, BP's U.S. refineries commonly provide repeated opportunities for employees to retake the same tests if needed. Some employees suggested that these repeat opportunities were almost unlimited. (Baker et al. 2007: 164)

Training for supervisors

One of the critical links in the management chain at Texas City was the front-line supervisors, described as *first level leaders*. The first level leaders were the first level of supervision for refinery operators; they worked close to the refinery hazards and in turn supervised the activities of operators, who also worked close to refinery hazards:

> Corporate managers emphasized the importance of first level leaders, stating that such a position is "the single most important position" and that the first level leader "holds it all together." (Baker et al. 2007: 154)

The Baker Panel looked at the training of first level supervisors across BP's North American operations and found:

> nothing to indicate that any of the refineries had a requirement that a new first level leader be tested on, or be familiar with, the equipment in the unit or that he or she have any minimum job tenure before promotion to that position. (Baker et al. 2007: 154)

Safety culture research suggests that front line supervision plays a critical role in the development of effective safety cultures (O'Dea 2003).

Lord Cullen in his inquiry into the Ladbroke Grove Rail disaster in October 1999 observed that:

> [Front line] companies in the rail industry should be expected to demonstrate that they have, and implement, a system to ensure that senior management spend an adequate time devoted to safety issues, with front line workers. Companies should make their own judgement on how much time their leaders should spend in the field, but best practice suggests at least one hour per week should be formally scheduled in the diaries of senior executives for this task. Middle ranking managers should have one hour per day devoted to it, **and first line managers should spend at least 30% of their time in the field.** (Cullen 2001: 64–65)

Given the central importance that front line supervision plays in ensuring a safe workplace, it is critical that supervisors are equipped with all the necessary tools and skills to adequately perform their roles.

Piper Alpha

At about 10 pm on 6 July 1988, an explosion erupted on the Piper Alpha platform, an oil and gas production facility in the North Sea. The explosion and subsequent inferno destroyed the platform. Of the 226 personnel on board, 165 were killed and a further two men died on board a rescue vessel. The disaster remains the world's worst offshore accident.

Like so many major accident events, the causes of the accident were many, varied, complex, and interrelated. In considering training and competence, we will look at just two of them: the permit to work system and the training in emergency response.

Permit to work

A permit to work (PTW) system is a formal, written, and documented system that is used to control certain types of high-risk work. On an oil and gas producing facility, work that might generate potential sources of ignition, such as welding, grinding, or other "hot work", is typically managed using a PTW. The PTW is designed to ensure that the potentially dangerous activity is well planned and supervised, and that other activities are not carried out nearby or at the same time in a way that might increase risk.

On Piper Alpha on the morning of July 6, a safety pressure valve was removed from one of two condensate pumps, Pump A, for routine maintenance, and the now open condensate pipe was temporarily sealed with a blind flange (flat metal disc); but the work could not be completed by the end of the shift so the blind flange remained in place.

The on-duty engineer filled out a PTW which stated that Pump A was not ready and must not be switched on under any circumstances.

When the night shift started, the PTW was left in the control room, but no one on the night shift was actually told about the condition of Pump A. Instead he had simply placed the permit in the control centre and left. At about 9:45 pm, the second condensate pump, Pump B, stopped suddenly and could not be restarted. This pump was critical, and the manager had only a few minutes to get the pump working again; otherwise, the power supply would fail completely. Although a search was made through the documents to see if Pump A could be started, the PTW stating that Pump A must not be started could not be found. No one doing the work knew that a critical part of the machine had been removed.

The manager assumed from the existing documents that it would be safe to start Pump A.

At about 9:55 pm, when Pump A was switched on, gas flowed into the pump, and because of the missing safety valve leaked out at high pressure, attracting the attention of workers and triggering gas alarms; but before anyone could act, the gas ignited and exploded.

The inquiry into the Piper Alpha disaster identified that *"in a number of significant respects"* the PTW procedure *"was habitually or frequently departed from"* (Cullen 1990: 191).

The safe execution of the PTW system was largely dependent on those managers who acted as Designated and Performing Authorities. Lord Cullen found that personnel appointed to act as Designated Authorities received no formal training at all, but rather were required to pick up their knowledge on the job while watching others perform the role. While recognising the role of "on the job" training, Lord Cullen also identified a potential weakness in over-reliance on the job training, namely that it *"suffers from the crucial weakness of perpetuating or accumulating errors"* (Cullen 1990: 194).

The PTW system was also required to be operated by a number of contractors engaged on the Piper Alpha, but no formal training was provided to these contractors—the responsibility lay with the contractors themselves:

> According to Mr A C B Todd, maintenance superintendent, under whose authority maintenance contractors work, [they] organised no training for contractors' employees in regard to the [PTW] system. In his view the long-term contractors would

be familiar with the system. ... when Mr Rankin
came to [Mr Todd's] office on 28 June he asked if
he knew the PTW system. Mr Rankin said he was
happy with it and knew how to work it. Mr Todd
did not probe to determine whether this was the
case. (Cullen 1990: 194)

Ultimately, Lord Cullen found:

I consider that the training required to ensure an
effective permit to work system was operated in
practice was not provided. (Cullen 1990: 195)

The failure within the training systems were compounded by the
inadequacies in reviewing the system—another very strong example of
the gap between the system as imagined and the system as it actually
is. One of the line managers, Mr Richards, recognised that he had an
obligation to maintain and monitor the safe systems of work, including
the PTW, but his "assurance" was largely passive: if problems were not
brought to his attention, he assumed that it was being implemented effectively. Ultimately *"he was surprised at the number of deficiencies"* in the PTW
system and asked *"how those deficiencies could exist without his knowledge"*
(Cullen 1990: 230).

The managers who had responsibility for the correct
operation of the PTW system were all aware that
the safety personnel on the platform were expected
to monitor the daily operation of the system. All
of them assumed that because they received no
reports of failings the system was working properly. However, none of them checked the quality of
that monitoring nor did they carry out more than
the most cursory examination of permits when they
had occasion to visit Piper.

...

The evidence ... shows, in my view, the operation of the PTW system was not being adequately
monitored or audited. These were failures for which
management were responsible. If there had been
adequate monitoring and auditing it is likely that
these deficiencies in the PTW system would have
been corrected. (Cullen 1990: 213)

Training in emergency response

Lord Cullen recognised, however, that the importance of emergency response and training had been identified in the case of Piper Alpha:

> I am not satisfied that the system as operated by Piper Alpha came close to achieving the necessary understanding on the part of all personnel and to how to react in the event of an emergency. (Cullen 1990: 241)

The errors identified ranged across all aspects of the emergency response systems, including induction processes, information contained in handbooks, and the conduct of emergency drills. In one example, a handbook provided to new-starters, which stated *"study it well—it may be your passport to survival"*, contained factual errors, referring to a method of throwing life raft capsules that did not apply on Piper Alpha and a reference to using scramble nets, which had been removed from Piper Alpha in the early 1980s (Cullen 1990: 211).

The evidence of the survivors of the disaster also painted a damming picture of the induction processes applied on Piper Alpha:

> 26 of the survivors (all contractors' personnel) were asked whether they had received a safety induction. Six of them said they had never done so. One thought that he had not: and one could not recall. The remaining 18 said that they had received an induction. But 4 said that it had lasted for 5–10 minutes. (Cullen 1990: 214)

Mr Patience, a lead safety operator, had given evidence that the inductions usually lasted 45 minutes (Cullen 1990: 211).

One worker who arrived on Piper Alpha 2 days before the disaster, when asked about his training, gave the following evidence:

> "He asked if we had been on the Piper before. I said 'No'. He said 'Have you worked offshore before?', and I said 'Yes'. He said 'Well you will know what the score is then'. That was much about what it was." (Cullen 1990: 213)

Another example was Mr Barton:

> He said: "When I was on this rig I was asked if I had worked on an offshore rig before, and I said 'Yes',

> that I had been on two, that I had been on Piper
> in 1982 and also the Claymore in 1985. This was at
> the safety induction, so he said 'Well, nothing has
> changed'." That was the sum total of the induction.
> (Cullen 1990: 213)

The Piper Alpha inquiry also identified gaps in the conduct of emergency drill and training: no full-scale emergency scenario training were conducted in the three years before the disaster; weekly specialist drill in areas such as fire-fighting, casualty handling, first-aid and other areas were not conducted at weekly intervals *"or anything approaching that"*; and evacuation drills, scheduled to be run once a week, were very limited (Cullen 1990: 214–215).

Once again, like the failure to ensure an adequate PTW system, the failure in emergency management was found to be a failure of management:

> The evidence to which I have referred above
> demonstrates that none of the drills required for
> practising evacuation procedures for the platform
> personnel or for the training of persons who had
> specific duties to perform in an emergency were
> carried out to the frequency predetermined by ...
> management. The responsibility for this failure lay
> with the platform management and the OIMs in
> particular. In my view they did not demonstrate
> the necessary determination to ensure that regu-
> latory was achieved or dissatisfaction expressed
> with the inadequate results. The lack of a deter-
> mined commitment to emergency training on the
> platform meant that the platform personnel were
> not as prepared for the disaster as they should
> have been. While the platform management did
> not exhibit the leadership required in this impor-
> tant area of training, the onshore safety staff did
> not operate an effective monitoring system with
> regard to emergency training. Where strong criti-
> cal comment was called for they were ineffective.
> (Cullen 1990: 218)

Deepwater Horizon

There were numerous examples in the Deepwater Horizon disaster that reflect the types of failures in training identified in earlier disasters.

Examples include the lack of guidance on safety-critical processes, operators being overwhelmed by the situations that they faced, critical emergency processes not being activated at the first available opportunity, and emergency response crews generally unequipped to deal with the circumstances that they faced.

Again, two examples serve to illustrate the gaps in training that may have impacted on the disaster: the negative pressure test, and sounding the general alarm.

The negative pressure test

The negative pressure test was an important safety check in the work being undertaken on the Deepwater Horizon prior to the explosion. The test was designed to reduce the pressure inside the well to simulate conditions that would be in place after the Deepwater Horizon had moved off the well. Pressure increasing inside the well during the test or fluids flowing up from the well could indicate a problem with the integrity of the well, and further work would be needed to re-establish the well's integrity.

In the end, the negative pressure test did indicate a problem—but it was not identified, and the data that was received was explained away by reference to the "bladder effect" (Graham et al. 2011: 6).

> It is now undisputed that the negative-pressure test at Macondo was conducted and interpreted improperly.
>
> ...
>
> The pressure data were not ambiguous. Rather, they showed repeatedly that formation fluids, in this case hydrocarbons, were flowing into the well. The failure to properly conduct and interpret the negative-pressure test was a major contributing factor to the blowout.
>
> ...
>
> It appears instead they started from the assumption that the well could not be flowing, and kept running tests and coming up with various explanations until they had convinced themselves their assumption was correct.
>
> The Commission has identified a number of potential factors that may have contributed to the failure to properly conduct and interpret the negative pressure test that night:

- First, there was no standard procedure for running or interpreting the test in either MMS regulations or written industry protocols. Indeed, the regulations and standards did not require BP to run a negative-pressure test at all.
- Second, BP and Transocean had no internal procedures for running or interpreting negative-pressure tests, and had not formally trained their personnel in how to do so.
- Third, the BP Macondo team did not provide the Well Site Leaders or rig crew with specific procedures for performing the negative-pressure test at Macondo.
- Fourth, BP did not have in place (or did not enforce) any policy that would have required personnel to call back to shore for a second opinion about confusing data. (Graham et al. 2011: 119)

Sounding the general alarm

One of the many defences available on the Deepwater Horizon was a system of general alarms. Detectors on board the Deepwater Horizon were able to monitor for and detect potential hazards developing—hazards such as combustible gas. Once detected, a visible and audible alarm would sound on the bridge of the Deepwater Horizon. These alarms were monitored by Dynamic Positioning Officers, or DPOs. The DPOs would acknowledge the alarms and then contact the area affected to then determine whether or not to sound the general alarm, which would alert the entire rig of the potential problem.

Fleytas

Andrea Fleytas was the Transocean DPO on duty in charge of the alarm panel at the time of the blowout. Although she was monitoring and acknowledging multiple alarms, she did not sound the general alarm.

Mr Bickford: And do you have any training course looking at a screen with alarms on it? In your training, do you have a physical simulation of alarms coming up on screens?

Ms Fleytas: Yes. For the dynamic positioning, yes, we do go over alarms and stuff.

Mr Bickford: What about combustible gas sensor alarms and other similar alarms on the rig?

Ms Fleytas: That training is on the Deepwater Horizon.

Mr Bickford: And so at Kongsberg you get none of that training either in the basic or the advanced course, correct?

Ms Fleytas: No, sir.

Mr Bickford: Describe to me what training you get on the Deepwater Horizon, on-the-job training that involves the identification of combustible gas alarms?

Ms Fleytas: The steps you need to take to figure out the problem. We acknowledge the alarm. We notify all the people that need to be notified. I let the senior DPO know. And if there is a combustible gas in a space, if it's one alarm in a specific space, we call that space to make sure that everybody is out. And then we send somebody down there with a sniffer, an oxygen level sensor to go down and check the area. If there is nothing in that area then we call the ETs and the ETs go and fix that sensor.

Mr Bickford: If, in fact, there is more than one sensor that goes off, does that change the calculus of what you just told me?

Ms Fleytas: Yes, sir.

Mr Bickford: How does that change it?

Ms Fleytas: If it's a combustible gas, we treat it like a fire. We make sure that that area is completely clear and we sound the general alarm.

Mr Bickford: So when two or more sensors of combustible gas go off in adjacent areas, it's your charge to sound the general alarm?

Ms Fleytas: Right.

Mr Bickford: Is that policy written down somewhere? Do they give you that as a training manual?

Ms Fleytas: Not that I know.

Mr Bickford: And so that policy is only passed on to you as a result of on-the-job training verbally?

Ms Fleytas: Yes, that's part of my training. (Fleytas 2010: 52–54)

...

Mr Bickford: Ms. Fleytas, why didn't you signal immediately the general alarm when two of the sensors came up magenta on the combustible gas alarms?

Ms Fleytas: It was a lot to take in. There was a lot going on. And soon after, I went over and hit the alarms.

Mr Bickford: But you didn't do it immediately, correct?

Ms Fleytas: No, sir.

Mr Bickford: And, in fact, at the time there were, by your testimony, more than ten to 20 magenta combustible gas alarms going off?

Ms Fleytas: Correct. (Fleytas 2010: 65)

The difficulties faced by Ms Fleytas were also reflected in the response by her supervisor, Mr Keplinger.

Mr Penton: Had you ever drilled or been trained at that scenario?
Mr Keplinger: What, as far as, you know, what just happened?
Mr Penton: Of a contemporaneous, simultaneous activation of all of the detection zones on that rig?
Mr Keplinger: No.
Mr Penton: And so you had never read in any of your Kongsberg training manuals, you were not taught by Kongsberg nor on the job on what to do in that scenario?

...

Mr Keplinger: Sir, I don't think anybody was trained for the massive detectors that were going off that night.
Hon. Judge Andersen: Okay. And there's an answer to that, that there hadn't been specific proactive training assuming this really difficult situation. Correct?
Mr Keplinger: Yeah. No. (Keplinger 2010: 296–298)

Clearly, the role of the DPO was central to the safety management systems—the emergency response systems in particular—on the Deepwater Horizon, and yet they were not adequately trained:

> Transocean likewise did not adequately train or drill its dynamic positioning officers (DPOs) on how to respond to emergency situations. (Bartlit, Sankar and Grimsley 2011: 237)

Conclusion

There is one other element that could have been dealt with under the heading of training and competence, and it arises particularly from Piper Alpha and the Deepwater Horizon. It goes to the issue of the often professed right of employees to *"stop the job"*; however, in both Piper Alpha and the Deepwater Horizon there were examples of critical safety steps that, in the ordinary course of events, would have severely disrupted operations at a great cost to the respective organisations and that, even at the height of the disasters in question, workers, including senior management, seemed reluctant to take. In the case of Piper Alpha it was a failure by two separate facilities to shut off the flow of oil that was feeding the fire on Piper Alpha, and in the case of the Deepwater Horizon, there was a failure to activate the EDS—emergency disconnect switch—which was designed to disconnect the rig from the underwater infrastructure.

To some extent, these failures could be attributable to issues of training and competence. They may also be indicative of reluctance to compromise production, even in the face of looming catastrophe. In any event, the questions arising from this reluctance, spread across two disasters more than 20 years apart, make them worth separate consideration, which we will do in Chapter 6.

References

Baker, J. et al. 2007. *The report of the BP U.S. Refineries Independent Safety Review Panel*. U.S. Chemical Safety and Hazard Investigation Board, Washington. http://www.bp.com/liveassets/bp_internet/globalbp/globalbp_uk_english/reports_and_publications/presentations/STAGING/local_assets/pdf/Baker_panel_report.pdf.

Bartlit, F., S. Sankar and S. Grimsley. 2011. Chief Counsel's report. National Commission on the BP Deepwater Horizon Oil Spill and Offshore Drilling. http://www.oilspillcommission.gov/sites/default/files/documents/C21462-407_CCR_for_print_0.pdf (accessed 17 February 2011).

CSB (U.S. Chemical Safety and Hazard Investigation Board). 2007. *Investigation report: Refinery explosion and fire*. Washington. http://www.csb.gov/assets/document/CSBFinalReportBP.pdf (accessed 23 November 2010).

Cullen, Lord. 1990. *The public inquiry into the Piper Alpha disaster*. Department of Energy. London: HMSO.

Cullen, Lord. 2001. *The Ladbroke Grove rail inquiry*. Health and Safety Executive. London: HMSO.

Dawson, D. and B. Brooks. 1999. *Report of the Longford Royal Commission: The Esso Longford gas plant accident*. Melbourne: Government Printer for the State of Victoria.

Fleytas, A. 2010. *Transcript: Deepwater Horizon Joint Investigation*, 5 October 2010. http://www.deepwaterinvestigation.com/go/doctype/3043/56779/.

Graham, B. et al. 2011. *Deep water: The Gulf oil disaster and the future of offshore drilling*. Report to the President. National Commission on the BP Deepwater Horizon Oil Spill and Offshore Drilling. http://www.oilspillcommission.gov/final-report (accessed 11 January 2011).

Keplinger, Y. 2010. *Transcript: Deepwater Horizon Joint Investigation*, 5 October 2010. http://www.deepwaterinvestigation.com/go/doctype/3043/56779/.

O'Dea, A. and R. Flin. 2003. The role of managerial leadership in determining workplace safety outcomes. Suffolk: HSE Books.

chapter 6

Everyone has the right to stop the job

"How do you know it's bad enough to act fast?"[*]

Introduction

In many jurisdictions the right of workers to stop work when they believe that there is a risk to their safety and health is enshrined in safety legislation, although in some cases it is not a general right to stop work, but requires some threat to the employees' safety at the time. For example, safety and health legislation in Western Australia provides that an employee can refuse to work if *"he or she has* **reasonable grounds** *to believe that to continue to work would expose him or her or any other person to a risk of* **imminent** *and serious injury or* **imminent** *and serious harm to his or her health."*[†] [emphasis added].

What is in many jurisdictions a legal right is often adopted and expanded as a matter of policy and a tool of safety management in many organisations, expanded in the sense that workers have an unqualified right to stop work if they have safety concerns—their right to stop work is not contingent on conditions such as a *reasonable belief* or *imminent* threat or risk. Certainly a policy allowing workers to stop work if they felt that conditions were unsafe was in effect at BP at the time of the Deepwater Horizon disaster:

> Mr Hayward: It is a policy that is real. And if anyone at any time believes that what they're doing is unsafe, they have both the right and the obligation to stop the task. (Hayward 2010: 141)

However, like so many of the elements examined in this book, a pronouncement by management that workers have a right and an obligation to stop work if they feel it is unsafe is no assurance that they will do it, and no assurance that the risks are actually being controlled. Indeed a

---

[*] Fred Bartlit, chief counsel for the National Commission on the BP Deepwater Horizon Oil Spill and Offshore Drilling

[†] Occupational Safety & Health Act 1984 (WA) section 26(1)

genuinely held belief by managers—genuinely held but untested and unverified—that workers will stop work if they feel it is unsafe can contribute to the illusion of safety; it is the gap between the system as it is imagined by management and the system as actually implemented in the workplace.

What assurance does management have, first, that workers have the skills and knowledge to recognise when they need to stop work, and how confident is the workforce in management's commitment to safety to actually exercise their "right" to stop the work?

Woods (2008: 31) poses the following scenario:

> Someone noticed there might be a problem developing, but the evidence is subtle or ambiguous. This person has the courage to speak up and stop the production process underway. After the aircraft gets back on the ground or after the system is dismantled or after the hint is chased down with additional data, they all discover the courageous voice was correct. There was a problem that would otherwise have been missed and to have continued would have resulted in failure, losses, and injuries. The story closes with an image of accolades for the courageous voice.
>
> When the speaker finishes the story, the audience sighs with appreciation—that was an admirable voice and it illustrates how a great organization encourages people to speak up about potential safety problems. You can almost see people in the audience thinking, "I wish my organization had a culture that helped people act this way."
>
> But this common story line has the wrong ending. It is a quite different ending that provides the true test for a high resilience organization.
>
> When they go look, after the landing or after dismantling the device or after the extra tests were run, everything turns out to be OK. The evidence of a problem isn't there or may be ambiguous; production apparently did not need to be stopped. Now, how does the organization's management react? How do the courageous voice's peers react?
>
> For there to be high resilience, the organization has to recognize the voice as courageous and valuable even though the result was apparently an unnecessary sacrifice on production and efficiency

goals. Otherwise, people balancing multiple goals will tend to act riskier than we want them to, or riskier than they themselves really want to.

How do you know it is bad enough to act?

One of the most important assumptions underlying any safety management strategies based on workers' right to stop work that they feel is unsafe is an assumption that the workers have a sufficient understanding of the risks in the business to know that they are involved in something that is unsafe.

On this level, the notion that everybody has a right to stop work that they believe to be unsafe is closely associated with training and competence:

- If the relevant operators on board the Deepwater Horizon had recognized or understood the implications of the failed negative pressure test, they may have had the opportunity to stop the job and reconsider the consequences of moving forward.
- If the relevant operators on board the West Atlas had recognized or understood the implications of pumping back 16.5 barrels of fluid and the potential for displacing the cement shoe, they may have had the opportunity to stop the job and seek further verification and confirmation that they had in fact created a competent barrier.
- If the Gas Plant operators at Longford on 25 September 1998 had understood the risks associated with the loss of lean oil flow, then they might have been in a position to stop the process and seek advice.
- If the operators at Texas City had understood the consequences of overfilling the raffinate splitter tower or had reasonable access to information to tell them how over-filled the tower was, then they may have been in a position to stop the job and recover the process.

In the end, any right to stop the job, not matter how fully and genuinely supported by management, would have been unlikely to influence the outcome because there is nothing to suggest that workers involved in these incidents had the knowledge necessary to recognise that they needed to stop the job.

Stopping the oil on Piper Alpha

The Piper Alpha was connected to two other oil-producing platforms, Claymore and Tartan. The Piper Alpha Inquiry determined that the fire on the Piper Alpha was not fuelled solely by the inventory of oil on the rig at the time of the explosion, but most likely was fed by inventories of oil that continued to be pumped from the Claymore and Tartan platforms (Cullen 1990: 138).

The first explosion triggered in the Piper Alpha disaster occurred at about 2155 hours on 6 July 1988.

The Claymore platform continued to produce oil until 2310 hours. A controlled shutdown of production had commenced on Claymore at about 2300 hours but they did not initiate an emergency shut down, which would have taken immediate effect (Cullen 1990: 138).

On Tartan, steps were taken to shut down production between 2230 and 2245 hours, with the final steps being taken at 2352 hours. Once again, an emergency shutdown, taking immediate effect, was not initiated (Cullen 1990: 138).

How is it that emergency steps that could have been taken to lessen the impact of the disaster on Piper Alpha were not initiated?

Claymore

When the offshore installation manager (OIM) was first told about the fire and explosion on the Piper Alpha at 2200 hours, he could not see it from the Claymore and thought that it would be controlled on Piper Alpha. At about 2215 the OIM was told by one of his managers that he wanted to shut down production because of the risk of oil being released onto the Piper Alpha. The OIM checked and confirmed that pressure in the pipelines was stable and decided to continue production. Between 2230 and about 2250 there were three requests to shut down production, but the OIM continued to maintain production—he did not think that the position on the Piper Alpha was beyond the control of the vessels' fire pumps.

By about 2300, following reports of a second explosion on Piper Alpha, the OIM realised that the situation was beyond Piper Alpha's ability to control and commenced to shut down production.

Tartan

After hearing of the situation of Piper Alpha, the OIM on Tartan was able to see the flames from the facility, 12 miles away. The OIM gave a direction to the production supervisor, Mr Moreton, to monitor the pressure in the gas pipeline. At about 2225 hours:

> Mr Moreton was told of a large explosion on Piper. He looked in the direction of Piper and saw a fireball. He had noticed that there had been a sharp drop in the pressure of the gas pipeline between [2220 and 2225] hours. He thought this odd, discussed it

> with someone else but could not explain it. He did
> not associate it with the large explosion on Piper
> although "it is apparent now".
>
> Both Mr Moreton and the OIM seemed to
> underestimate the impact of what was occurring on
> Piper Alpha, both assuming that the situation could
> be controlled on Piper Alpha. (Cullen 1990: 142)

In the end, Lord Cullen found that production could have been shut down earlier on both Claymore and Tartan, and that there was authority to do so, but that it may have been an event which the personnel involved were ill equipped to respond to:

> The strong impression with which I am left after
> hearing the evidence as to the response of Claymore
> and Tartan was that the type of emergency with
> which the senior personnel of each platform was
> confronted was something for which they had not
> been prepared. (Cullen 1990: 144)

The scenario with which management were faced was one that had never been discussed or considered, much less rehearsed.

In the case of Claymore, there was an added difficulty; the Piper Alpha Inquiry determined that the OIM was:

> reluctant to take responsibility for shutting down
> oil production. (Cullen 1990: 144)

Activating the EDS on the Deepwater Horizon

One of the many safety barriers on the Deepwater Horizon was the Emergency Disconnect System, the EDS. The effect of the EDS, as the name implies, was to "disconnect" the Deepwater Horizon from the well in the event of an emergency. The EDS should have closed a device, the blind shear ram, which should have severed the drill pipe, sealed the well, and disconnected the rig from the BOP. In the event, the EDS did not work, possibly because some of the control panels were damaged in the first explosion.

Whatever the technical reasons for the failure of the EDS, there was another underlying weakness in relying on the EDS as a safety barrier: the ability and willingness of people to make the *"momentous decision"* (Graham et al. 2011: 114) to activate it.

By now, Winslow began to wonder why the derrick was still roaring with flames. Hadn't the blowout preventer been activated, sealing off the well and thus cutting off fuel for the conflagration? He headed to the bridge. Kuchta said, "We've got no power, we've got no water, no emergency generator."

Steve Bertone was still at his station on the bridge and he noticed Christopher Pleasant, one of the subsea engineers, standing next to the panel with the emergency disconnect switch (EDS) to the blowout preventer.

Bertone hollered to Pleasant: "Have you EDSed?"

Pleasant replied he needed permission. Bertone asked Winslow was it okay and Winslow said yes.

Somebody on the bridge yelled, "He cannot EDS without the OIM's [offshore installation manager's] approval."

Harrell, still dazed, somewhat blinded and deafened, had also made it to the bridge, as had BP's Vidrine. With the rig still "latched" to the Macondo well, Harrell was in charge. Bertone yelled, "Can we EDS?" and Harrell yelled back, "Yes, EDS, EDS."

Pleasant opened the clear door covering the panel and pushed the button.

Bertone: "I need confirmation that we have EDSed."

Pleasant: "Yes, we've EDSed."

Bertone: "Chris, I need confirmation again. Have we EDSed?"

Pleasant: "Yes."

Bertone: "Chris, I have to be certain. Have we EDSed?"

Pleasant: "Yes." He pointed to a light in the panel.

(Graham et al. 2011: 1314)

Whatever aspirations an organisation might have about its safety management system, and whatever genuinely held beliefs that senior managers might have that workers have the right to stop work if they believe that anything is unsafe, it is one thing to say you have a right to stop work, it is another thing altogether to know when to exercise that right, and then to execute it.

As the National Commission rightly identified:

> Finally, and perhaps most significantly, the rig crew had not been trained adequately how to respond to such an emergency situation. In the future, well-control training should include simulations and drills for such emergencies—including the momentous decision to engage the blind shear rams or trigger the EDS. (Graham et al. 2011: 122)

Conclusion

I had always found the comparisons between Piper Alpha and the Deepwater Horizon striking: that seemingly competent people, in the face of obvious and escalating disaster, could be so reluctant to initiate important safety measures. Without knowing precisely what was in the heads of the people caught up in the frightening events at the time, it seems to me to speak volumes of the cultures that had been built up in the organisations and the fragility of a safety management system that relies on individuals to both recognise a dangerous situation and then have the courage to act on it.

While writing this book, I was on a flight from Western Australia to South Australia and was re-reading *Lesson from Longford* by Andrew Hopkins (2001) when I was struck by a single extract of evidence in the Longford Royal Commission, evidence that could have been taken from the Deepwater Horizon or Piper Alpha, when one operator said:

> I would go so far as to say I faced a dilemma on the day, standing 20 metres from the explosion and the fire, as to whether or not I should activate ESD 1 (Emergency Shutdown 1), because I was, for some strange reason, worried about the possible impact on production. (Hopkins 2001: 146)

I cannot help but wonder what it is that gives managers and organisations the confidence and assurance that any message that "everybody has the right to stop the job", no matter how truly held, is a message that has real resonance and meaning where it counts.

References

Cullen, Lord. 1990. *The public inquiry into the Piper Alpha disaster*. Department of Energy. London: HMSO.

Graham, B. et al. 2011. *Deep water: The Gulf oil disaster and the future of offshore drilling.* Report to the President. National Commission on the BP Deepwater Horizon Oil Spill and Offshore Drilling. http://www.oilspillcommission.gov/final-report (accessed 11 January 2011).

Hayward, T. 2010. Transcript: U.S. House of Representatives, Subcommittee on Oversight and Investigations, Committee on Energy and Commerce. The role of BP in the Deepwater Horizon oil spill, 17 June 2010. Washington, D.C. http://energycommerce.house.gov/documents/20100617/transcript.06.17.2010.oi.pdf (accessed 23 November 2010).

Hopkins, A. 2001. *Lessons from Longford: The Esso gas plant explosion.* Sydney: CCH Australia Limited.

Woods, D. 2008. Essential characteristics of resilience. In Hollnagel, E., D. Woods and N. Leveson. *Resilience engineering concepts and precepts.* England: Ashgate.

chapter 7

Delegation

"He had a responsibility not only to give instructions, but also to see to it that those instructions were carried out in order to minimize the damage."[*]

Introduction

I have called this chapter "Delegation" as this is a common and well-understood way of talking about having "other people" carry out our responsibilities, but the chapter is really about reliance on other people and the extent to which we can rely on others to discharge our responsibilities for safety and health.

Delegation is a normal and entirely proper business practice. It is normal and entirely proper in the same way that businesses legitimately pursue efficiency in order to expand and grow, or in some cases to survive. However, like efficiency, delegation needs to be understood in the context of its impact on safety and in the context of a manager understanding how those delegated functions are carried out.

One of the common and problematic areas of delegation in safety management is contractor safety management. However, this book is not directly about contractor safety management, although the principles discussed in this chapter are as relevant to delegation to contractors as they are to delegating to individuals of work groups. How do you know that the people that you are delegating responsibilities to have the skills, experience, training, resources, or other requirements to meet the expectations in relation to safety and health, and how do you ensure that those expectations are being met?

In the Bata decisions the balance between effective risk management and delegation was brought into sharp relief in the examination of Mr Marchant:

> In the circumstances, it is my opinion that due diligence requires him to exercise a degree of supervision and control that "demonstrate that he was exhorting those whom he may normally be expected to influence or control to an accepted standard of behaviour."

[*] R v Bata Industries Ltd 7 C.E.L.R. (N.S.) 245, 9 O.R. (3d) 329 (Bata), [164–165]

He had a responsibility not only to give instructions, but also to see to it that those instructions
were carried out in order to minimize the damage.*

And further in the case of Mr Weston:

He had an obligation if he decided to delegate
responsibility, to ensure that the delegate received
the training necessary for the job, and to receive
detailed reports from that delegate.†

This juxtaposition between delegation and understanding risk
was also evident in Tony Hayward's evidence before the US House
of Representatives Subcommittee on Oversight and Investigations,
Committee on Energy and Commerce looking at the role of BP in the
Deepwater Horizon explosion and oil spill—for example, when asked
questions about the decision to choose the long string design:

Mr Doyle: So we have reviewed all of their emails and communications. We find no record that they knew anything about this decision. In fact, we find no evidence that they ever received briefings
on the activities aboard the Deepwater Horizon before the explosion. These decisions all seem to have been delegated to much lower
ranking officials.
*Well, Mr Hayward, then, who was the one who made the decision to use a single tube
of metal from the top of the well to the bottom? Who did make that decision?*
Mr Hayward: I am not sure exactly who made the decision. It would have
been a decision taken by the drilling organisation in the Gulf of
Mexico. They are the technical experts that have the technical knowledge. (Hayward 2010: 128–129)

It is difficult to criticize this position in and of itself—senior managers regularly and properly rely on other managers and technical experts
to make business decisions every day. To suggest that an organisation the
size of BP (or even much, much smaller businesses) could operate with
close CEO or senior executive involvement in day-to-day or even more
strategic business decisions is clearly unrealistic. But the disconnect
between delegation and ensuring that safety risks are controlled becomes
evident in the subsequent (seemingly) frustrated comment:

* Bata, [164–165]
† Bata, [172–173]

> And now we are all paying the consequences
> because those of you at the top don't seem to have a
> clue about what was going on on this rig. (Hayward
> 2010: 130)

How is it that managers, in the entirely proper reliance on others often more technically experienced than they are or closer to the day-to-day operations of the business, get assurance that risk is being properly managed?

Commercial Industrial Construction Group[*]

CICG was a construction company doing building work on a four-storey building. Part of the work involved the removal of a section of the first floor roof. During the work the site manager, Peter Bacon, directed George Podger, a plasterer/carpenter with no scaffolding skills or experience, to erect some scaffolding beneath the roof that was to be removed. The scaffolding was later inspected by government inspectors and found to be unsafe.

The day after the scaffolding had been erected Bacon directed Jason Roach, a builder's labourer, who had been employed by CICG for about a year, to remove a section of the roof on the first floor. Bacon and Roach discussed the need for a Job Safety Analysis (JSA), a written document to identify and manage hazards associated with the work, but Bacon told Roach that a JSA was required for the job and no JSA was prepared.

When Podger arrived at the site, Bacon directed him to help Roach with removing the section of roof.

Roach could not access part of the roof from the scaffolding and was working directly on the roof when he stepped over an opening created by the removal of some of the iron sheeting. As he did so, his foot slipped on the edge of the opening and he fell through about three metres onto the concrete floor below. Roach suffered bruising and lacerations to his foot, arms and back, and from shock. He was off work for a week.

The day after the fall Bacon directed Podger to go back to the roof and finish the job. He did not give Podger any safety instructions or put in place any additional safety procedures or fall protection.

Following Union and government safety regulator visits to the site, CICG was charged and pleaded guilty to charges under safety and health legislation that they did not:

> provide and maintain so far as was practicable for
> employees a working environment that was safe
> and without risks to health in that [it] failed to pro-
> vide and maintain systems of work in relation to a

[*] R v Commercial Industrial Construction Group Pty Ltd [2006] VSCA 181, [45] (CICG)

> construction site … that were so far as was practicable safe and without risks to health.[*]

Although pleading guilty, CICG sought to minimise the penalty imposed by the Court, and argued that the seriousness of its breach was lessened by the fact that:

> the failure in this case was that of an individual who was actually charged with ensuring that the safety management system was properly carried out.[†]

The Court paid little regard to that position.

> We are wholly unpersuaded by this argument. For any employee to behave as Bacon did, with blatant disregard for the safety of his fellow employees, would be bad enough. But when the employee in question is the person with supervisory responsibilities, including responsibility for ensuring safety at the site, the gravity of the company's breach is increased, not reduced. **It is difficult to understand how the company could have allowed someone with Bacon's apparent indifference to risk to occupy such a position.**[‡] [emphasis added]

Earlier we looked at the position of front line supervision in the Texas City refinery explosion, noting the need to ensure that front line supervision have all of the skills and tools necessary to fulfil their critical role in effective safety management. What the CICG case reminds us of is the importance of not only ensuring that those skills and tools are in place, but also that they are effectively implemented.

HMAS Westralia

On Tuesday, 5 May 1998, four Royal Australian Navy Personnel died as a result of a major fire in the engine room of the ship *HMAS Westralia*, which started when flexible fuel hoses failed, providing a source of diesel fuel, which ignited. The flexible fuel hoses had been installed to replace old ones as part of a refurbishment of the vessel's metal pipes and had been in operation for only a little over 36 hours at the time of the fire.

[*] Ibid, [3]
[†] Ibid, [42]
[‡] Ibid, [43]

A coronial inquiry into the fatalities identified that no engineering assessment had been conducted on the flexible fuel hoses, which was a serious failure resulting from a series of mistakes and the system's deficiencies.

In the context of delegation and relying on others to carry out the roles relevant to health and safety, the Coroner found that the contract manager, Warrant Officer Jones, was not appropriately skilled to manage the contract on the scale and range in question. The Coroner stated:

> Warrant Officer Jones did not have sufficient engineering skills to be able to identify potentially dangerous problems associated with the works. Warrant Officer Jones had completed a 6 month qualifying engineering course 20 years before the fire and that was his highest level of relevant technical training.
>
> The Ordering Authority should have ensured that there was adequately skilled supervision of the contract to ensure that the contract was addressing any safety issues which could arise. This was particularly the case because all of the ship's repair and maintenance requirements were being outsourced ... Warrant Officer Jones did not have sufficient relevant knowledge to identify complex safety issues and had received no training as to the Lloyd's requirements for the ship. (Western Australian State Coroner 2003: 9)

In view of the evidence before the coroner, it was accepted that there was no audit process in place to identify errors in procedures and it was assumed that Warrant Officer Jones would do his job correctly without any independent monitoring (Western Australian State Coroner 2003: 73).

While less dramatic than the CICG case, the principles are still the same: how do managers have assurance that the people with safety-critical roles are capable and are, in fact, effectively delivering on safety?

*The Ritchie Decision**

In January 2003 an employee of Owens Container Services Australia Pty Ltd was cleaning a tank in the company's tank wash facilities when there was an explosion and the worker was killed. The worker had sprayed methyl ethyl ketone (MEK) into the tank to soften some resin that needed

---

* Inspector Ken Kumar v David Aylmer Ritchie (2006) NSWIRComm 323 (Ritchie)

to be removed. MEK is a highly volatile and highly flammable substance that was being used as a cleaning agent.

After the tank had been sprayed with MEK, it was left for 20 to 30 minutes. When the employee returned to the job he used a high-pressure water spray gun to try to clean off the remaining resin.

Shortly after the explosion happened.

Although the cause of the explosion was not clear, the most likely explanation for it was that the water spray gun used by the employee generated static electricity, which ignited the MEK.

The CEO of the Owens Group, which included Owens Container Services, was Mr Ritchie. The Owens Group was a public company that owned 30 companies and operated in a number of countries, with 80 sites worldwide and employing around approximately 1600 staff. Mr Ritchie had been the CEO since October 2001.

As a consequence of the accident Mr Ritchie was charged and convicted of offences under New South Wales health and safety legislation.[*] In essence, the case against Mr Ritchie alleged that he did not use all due diligence to ensure that the workplace was safe. The Court found that the evidence did not disclose *"a director's mind concentrated on the risks of this operation"*, and that the failures evident in the accident *"dramatically demonstrate the absence of a safe system of work"* and *"the absence of due diligence by Mr Ritchie as a director of the company to address and correct those deficiencies ..."*[+]

> The defence was not advanced by the defendant emphasising how busy he was in the work of the Group, his geographical remoteness and his lack of daily involvement in the day-to-day operations of the business; precisely those factors made it imperative that the system he put in place or oversaw was proper and adequate to ensure compliance with the Act and that the means of ensuring the system was in force.
>
> ...
>
> Those administering the system had no means of effectively enforcing it and there was no evidence as to how the enforcement was achieved.
>
> ...
>
> The defendant did nothing to ensure that persons employed as occupational health and safety

[*] The Occupational Health and Safety Act 2000
[+] Ritchie, [177]

officers were trained. There was nothing in the system that would bring the lack of training of these people to the defendant's attention. There were significant matters that the defendant had no knowledge of but he had made assumptions.

...

The defendant's evidence, ultimately, relied upon a series of assumptions and reports he received from managers as well as audits in order to ensure the safety of the Race workforce. **He assumed that the managers were doing their job.**[*] [emphasis added]

Conclusion

There is a provoking question that was asked during the cross examination of PTT chief operating officer Andy Jacob during the Montara inquiry:

So whose responsibility was it in relation to ensuring that the well construction department was actually administering its affairs in accordance with PTT's expectations … ? (Jacob 2010: 1891)

This is the issue that goes to the heart of our reliance on others when it comes to health and safety obligations. How do we know the people we rely on to meet the business objectives are also meeting our expectations for safety and health?

References

Hayward, T. 2010. Transcript: U.S. House of Representatives, Subcommittee on Oversight and Investigations, Committee on Energy and Commerce. The role of BP in the Deepwater Horizon oil spill, 17 June 2010. Washington, D.C. http://energycommerce.house.gov/documents/20100617/transcript.06.17.2010.oi.pdf (accessed 23 November 2010).

Jacob, A. 2010. Transcript: Montara Commission of Inquiry. http://www.montarainquiry.gov.au/transcripts.html (accessed 29 September 2010).

Western Australian State Coroner. 2003. Inquest into the deaths of Shaun Damian Smith; Phillip John Carroll; Megan Anne Pelly and Bradley John Meek. http://www.navy.gov.au/w/images/HMAS_Westralia_Finding.pdf (accessed 20 November 2010).

---

[*] Ibid, [154–157]

chapter 8

Warning signs

"One of the significant matters which the interim report demonstrated was the fact a number of senior managers of the SRA were informed of the inherent deficiency in the deadman foot pedal, but despite many warnings no steps were taken to adequately test the device. This lack of action is difficult to understand."[*]

Introduction

Even a superficial review of the history of major accidents makes it clear that the precursors to disaster are not hidden—they are generally well known within an organisation, or parts of it. Major inquiries and legal proceedings all too often identify that clear warning signs about factors that cause or contribute to major accidents existed prior to the accident, and in many cases were brought to the attention of relevant managers.

It is often thought that warning signs are subtle—emerging gradually over time is a way that hides their true meaning or potential consequences. Sometimes, this may be true, and warning signs need to be proactively sought out in an effort to learn and improve. The types of processes that help organisations to learn include such things as:

- Understanding the "near miss" event
- Developing a culture that encourages incident and hazard reporting
- High-quality incident investigations
- Open sharing of lessons across and between parts of a business
- Learning from other industries and businesses

Often, however, these warning signs are direct and unambiguous— as unequivocal as we will likely *"kill someone in the next 12–18 months"* (CSB 2007: 25).

The reasons why these warnings signs are not responded to in a way that prevents a major accident are many and varied. Sometimes, warning signs are misunderstood because of the characteristics of individuals

[*] McInerney, P. QC. 2003: 215

who may lack the experience, skills, or competence to recognise warning signs for what they are and be able to respond appropriately. On other occasions there may be cultural or organisational factors that cause individuals to downplay or dismiss the importance of warning signs, factors such as the need to minimise costs, maximize production, or stick to a schedule—all of which can influence decision making so that responding to warning signs gets pushed to the background in the face of more pressing concerns.

Whatever the reason, in the cold hard light of a major workplace accident it is usually individual managers who find themselves faced with the difficult task of having to explain why they responded, or failed to respond, to warning signs as they did.

The expected management response to warning signs was well described in the Bata Decision, where, in reference to the quotation for the cost of environmental cleanup, the Court noted:

> It is my opinion, red flags should have been raised in [Mr Weston's] environmental consciousness when the first quote of $58,000 was obtained. Instead of simply dismissing it out of hand, he should have inquired why it was so high and investigated the problem. I find that he had no qualms about accepting the second quote of $28,000, and he had no further information other than it was cheaper.[*]

Red flags should have been raised and he should have made further enquiries.

Unfortunately, as history so clearly shows us, all too often red flags are not raised, no enquiries are made, and warning signs remain unattended until it is too late.

Peter Kite

Peter Kite was the managing director of OLL Ltd, trading under the name Active Learning and Leisure Ltd, a business that operated a leisure centre. On 22 March 1993, four school students aged 16 and 17 were drowned during a school open-sea canoeing trip whilst engaged in an activity at the centre. Eight students, two teachers and two instructors from OLL undertook the trip.

In June 1992, Mr Kite received a letter from two of his staff members. The letter stated:

---

[*] R v Bata Industries Ltd 7 C.E.L.R. (N.S.) 245, 9 O.R. (3d) 329 (Bata) at [170]

> At present we are walking a very fine line between "getting away with it" and having a very serious incident … We would also like to know why we do not get supplied with a first-aid kit and tow-line … It's unsafe and not organised … Having seen your 1993 brochure and planned expansion, we think you should have a very careful look at your standards of safety, otherwise you might find yourselves trying to explain why someone's son or daughter will not be coming home. Nobody wishes or wants that to happen, but it will soon or later.[*]

Unfortunately, these warnings were not appropriately responded to.

It was not suggested that Mr Kite was directly responsible for what occurred on 22 March 1993—he was not present. Ultimately, he was convicted and finally sentenced[†] to two years jail for manslaughter because of his failure to lay down a proper system to ensure safety.

The Herald of Free Enterprise

Prior to the disaster on 6 March 1987, there had been several instances of vessels owned by the company that operated the *Herald of Free Enterprise*, Townsend Car Ferries Limited, going to sea with either bow or stern doors opened. In June 1985 the captain of one of those vessels, Captain Blowers of the *Pride of Free Enterprise*, wrote a memorandum to Mr Develin, one of the directors of Townsend, which included the following:

> 4. Mimic Panel—There is no indication on the bridge as to whether the most important watertight doors are closed or not. That is the bow or stern doors. With the very short distance between the birth and the open sea on both sides of the Channel this can be a problem if the operator is delayed or having problems in closing the doors. Indicator lights on the very excellent mimic panel could enable the bridge team to monitor the situation in such circumstances. (Department of Transport 1987: 23)

[*] R v Kite (Peter Bayliss) (1996) 2 Cr. App. R. (S), 295, 297 (Kite)

[†] Mr Kite was originally sentenced to three years jail but the sentence was reduced to two years on appeal.

The response to the suggestion, particularly in light of the history of vessels sailing without bow or stern doors closed, was disconcerting to say the least:

> Mr Devlin circulated that memorandum amongst managers to comment. It was a serious memorandum which merited serious thought and tension, and called for a considered reply. The answers which Mr Devlin received will be set out verbatim. From Mr J. F. Alcindor, a deputy chief superintendent: "Do they need an indicator to tell from whether the deck storekeeper is awake and sober? My goodness!!" From Mr A. C. Reynolds: "Nice but don't we already pay someone!" From Mr R. Ellison: "Assume the guy who shuts the door tells the breach if there is a problem." From Mr D. R. Hamilton: "Nice!" It is hardly necessary for the Court to comment that these replies display an absence of any proper sense of responsibility. (Department of Transport 1987: 24)

The persistent nature of the failure to respond to warning signs leading up to the capsizing of the *Herald of Free Enterprise* was also evident in the way that the organisation addressed (or failed to address) concerns about the ability of the vessel Masters to determine the draught of their vessel before sailing. The inquiry identified that it was a legal requirement that the Master should know the draughts of their ship and that this had to be entered in the official log book before putting to sea. It was particularly important for the Master of the *Herald of Free Enterprise* to know the draught of the vessel because there were restrictions on the number of passengers that could be carried if the draught exceeded 5.5 meters.

Evidence in the inquiry established that no attempt had been made to read the draughts on a regular basis and that *fictitious figures** were entered in the Official Log. The inquiry found that:

> The difficulties faced by the Masters are exemplified by the attitude of Mr Develin to a memorandum dated 24th October 1983 and sent to him by Captain Martin.

---

* This failure is reminiscent of the matters discussed in the context of the Texas City refinery explosion, where valves were marked off as having been checked when they hadn't been.

The relevant passages of that memorandum are as follows:

> "For good order I feel I should acquaint you with some of the problems associated with one of the Spirit class ships operating to Zeebrugge using the single deck berths ...
>
> 4. At full speed, or even reduced speed, bow wave is above belting forward, and comes three quarters of the way up the bow door ...
>
> 6. Ship does not respond so well when trimmed so much by the head, and problems have been found when maneuvring ...
>
> 8. As you probably appreciate we never know how much cargo we are carrying, so that a situation could arise that not only are we overloaded by 400 tons but also trimmed by the head by 1.4 m. I have not been able to work out how that would affect our damage stability."

Mr Develin was asked what he thought of that memorandum. His answer was, "Initially I was not happy. When I studied it further, I decided it was an operational difficulty report and Captain Martin was acquainting me of it." Later he said, "I think if he had been unhappy with the problem he would have come in and banged my desk." When Mr Develin was asked what he thought about the information concerning the effect of full speed he said, "I believe he was exaggerating." In subsequent answers Mr Develin made it clear that he thought every complaint was an exaggeration. In reply to a further question Mr Develin said, "If he was that concerned he would not have sailed. I do not believe that there is anything wrong sailing with the vessel trimmed by the head." Mr Develin ought to have been alert to the serious effects of operating at large trims. Furthermore he should have been concerned about Captain Martin's remarks about stability. He should at least have checked the ship's stability book. If he had done so he would have found that the ship was operating outside her conditions as set out and, therefore, not complying with the conditions under

which the Passenger Ship Certificate was issued.
(Department of Transport 1987: 26)

A safety-related request, such as the request about the indicator lights in the *Herald of Free Enterprise* or the concerns about the sailing conditions, should not have been dismissed without well-considered, risk-based, and justifiable reasons. In the event of an accident, managers will be expected to account for their decision-making and to be able to demonstrate why they thought that their decision-making, in the face of any warning signs, was justified.

The Space Shuttle Challenger

A little over 12 months before the *Herald of Free Enterprise* disaster, on 28 January 1986 another tragedy rooted in the failure of management responses to warning signs occurred, the loss of the Space Shuttle *Challenger*.

The *Challenger* disintegrated 73 seconds into its flight, and resulted in the deaths of all 11 crewmembers.

Technically, the disaster happened because an O-ring seal in the right solid rocket booster failed at liftoff. The O-ring failure caused a breach in the solid rocket booster joint that it sealed, allowing pressurized gas to escape, leading to the structural failure of the external tank. Aerodynamic forces promptly broke up the orbiter.

The United States President, Ronald Reagan, appointed a special commission of inquiry, the Rogers Commission, to investigate the accident (Rogers et al. 1986).

The Rogers Commission found that NASA's organisational culture and decision-making processes had been a key contributing factor to the accident. NASA managers and its contractor, Morton Thiokol, had known that the design of the solid rocket boosters contained a potentially catastrophic flaw in the O-rings, but they failed to address it properly. They also disregarded warnings from engineers about the dangers low temperatures during the morning launch posed and had failed to adequately report these technical concerns to their superiors.

Findings

The genesis of the *Challenger* accident—the failure of the joint of the right solid rocket motor—began with decisions made in the design of the joint and in the failure by both Thiokol and NASA's Solid Rocket Booster Project Office to understand and respond to facts obtained during testing.

The Commission has concluded that neither Thiokol nor NASA responded adequately to internal warnings about the faulty seal design. Furthermore, Thiokol and NASA did not make a timely attempt to develop and verify a new seal after the initial design was shown to be deficient. Neither organization developed a solution to the unexpected occurrences of O-ring erosion and blow-by, even though this problem was experienced frequently during the Shuttle flight history. Instead, Thiokol and NASA management came to accept erosion and blow-by as unavoidable and an acceptable flight risk. Specifically, the Commission found that:

1. The joint test and certification program was inadequate. There was no requirement to configure the qualifications test motor as it would be in flight, and the motors were static tested in a horizontal position, not in the vertical flight position.
2. Prior to the accident, neither NASA nor Thiokol fully understood the mechanism by which the joint sealing action took place.
3. NASA and Thiokol accepted escalating risk apparently because they "got away with it last time". As Commissioner Feynman observed, the decision making was: "a kind of Russian roulette ... [The Shuttle] flies [with O-ring erosion] and nothing happens. Then it is suggested, therefore, that the risk is no longer so high for the next flights. We can lower our standards a little bit because we got away with it last time. ... You got away with it but it shouldn't be done over and over again like that."
4. NASA's system for tracking anomalies for Flight Readiness Reviews failed in that, despite a history of persistent O-ring erosion and blow-by, flight was still permitted. It failed again in the strange sequence of six consecutive launch constraint waivers prior to 51-L, permitting it to fly without any record of a waiver, or even of an explicit constraint. Tracking and continuing only anomalies that are "outside the data base" of prior flight allowed major problems to be removed from, and lost by, the reporting system.
5. The O-ring erosion history presented to Level I at NASA Headquarters in August 1985 was sufficiently detailed to require corrective action prior to the next flight.
6. A careful analysis of the flight history of O-ring performance would have revealed the correlation of O-ring damage and low temperature. Neither NASA nor Thiokol carried out such an analysis; consequently, they were unprepared to properly evaluate the risks of launching the 51-L mission in conditions more extreme than they had encountered before. (Rogers et al. 1986: 148)

Texas City

We have already looked at elements of the Texas City Refinery explosion in the context of both management line of sight and understanding the rules. But in addition to these matters there were a number of very clear warning signs given to management at all levels of BP of the perilous state of the safety performance at Texas City and the potential for the type of catastrophic event that eventually resulted.

The Texas City disaster provides a particular challenge in the context of "warning signs", in so much as it represents the full weight of a sophisticated safety management apparatus unearthing, articulating and promulgating very clear and unambiguous warnings about the potential for the type of failure that did occur, and yet the organisation being unable to respond in a way that addressed those warnings.

Audits, reviews, culture reviews, strategies, plans, accidents, including fatalities, all pointed to a strong risk of a serious accident. Warnings signs included a 2002 study that identified technical integrity at the Texas City refinery as *"one of the biggest problems"* (CSB 2007: 156), a 2003 audit that identified the poor condition of the infrastructure and assets at Texas City, a "checkbook mentality" that meant sufficient funds were not being applied to the problems (CSB 2007: 161–162).

Ultimately, these warning were not responded to:

> Beginning in 2002, BP Group and Texas City managers received numerous warning signals about a possible major catastrophe at Texas City. In particular, managers received warnings about serious deficiencies regarding the mechanical integrity of aging equipment, process safety, and the negative safety impacts of budget cuts and production pressures.
>
> However, BP Group oversight and Texas City management focused on personal safety rather than on process safety and preventing catastrophic incidents. Financial and personal safety metrics largely drove BP Group and Texas City performance, to the point that BP managers increased performance site bonuses even in the face of the three fatalities in 2004. Except for the 1,000 day goals, site business contracts, manager performance contracts, and VPP bonus metrics were unchanged as a result of the 2004 fatalities. (CSB 2007: 177–178)

Like the Peter Kite case discussed earlier, these warning were not hidden or subtle. The 2005 Texas City HSSE Business Plan warned that the refinery likely would *"kill someone in the next 12–18 months"* (CSB 2007: 177), and the HSE manager, referring to a catastrophic incident said, in early 2005, *"I truly believe that we are on the verge of something bigger happening"* (CSB 2007: 177).

Montara

One of the key themes to arise out of the Montara Inquiry was that there was a prevailing philosophy to get the job done without delay, a philosophy that resulted in cost minimization at the expense of safety and well integrity (Borthwick 2010: 11). This finding was supported not only in part by the conduct of the organisation as evidenced in the hearing, but also by evidence of warnings received prior to the accident, warnings that pointed directly towards such a culture.

During cross examination of PTT's well construction manager, Craig Duncan, there was a good deal of questioning about Mr Duncan's consideration of various correspondence and accusations made by a contractor, Advanced Well Technologies. In essence, Advanced Well Technologies had terminated their services with PTT because, from Advanced Well Technologies' perspective, PTT was continually putting "production before safety".

Mr Howe QC: Then in the middle of the next paragraph, they express deep concern with the potential for damage to their professional reputation:

… if [AWT] are to continue to be seen to be responsible for the completions design and implementation of this project whilst not able to influence (in our view) critical decisions.

That's a fairly damning position that they are taking with respect to PTT, isn't it?

Mr Duncan: Yes.

Mr Howe QC: They are, in effect, saying, "You are implicating us in planning and design of production in a way that we think is going to adversely affect our own professional standing and reputation"?

Mr Duncan: That's what they are saying, yes.

Mr Howe QC: And they didn't want to be a part of it?

Mr Duncan: That's right.

Mr Howe QC: They made reference to PTT pursuing cost savings over proper risk management; that's right, isn't it?

Mr Duncan: We certainly had a difference of opinion in that, yes.

Mr Howe QC: Then they go on in that same paragraph by saying, "We respect that it is PTT's project." They acknowledge, as it were, that all projects need a healthy cost focus and that, from time to time, PTT's management team would need to take project decisions which balance a variety of interests. Do you see that?

Mr Duncan: Yes.

Mr Howe QC: Then they offer something which might, on one view, have some prophetic resonance:

… we believe that a failure to assess and quantify the risks associated with such decisions will ultimately be to the detriment of the execution budget and longer term operability of the production system.

Do you see that?
Mr Duncan: Yes. (Duncan 2010: 1380–1391)

...

Mr Howe QC: Can I ask you, in the period after the time of this document, which was toward the end of 2007, did you generally continue
   to approach your balancing of time/cost savings on the one hand and
   attention to proper well engineering standards on the other hand in
   the same general way that had provoked AWT's criticisms?
Mr Duncan: I did tend to go for the lowest cost, fit-for-purpose solutions,
   which I think is appropriate.
Mr Howe QC: So the general sort of balancing between sometimes competing imperatives, which is addressed by AWT at this time, is something that you continued to adhere to in the lead-up to and in the
   course of events from March to August last year; is that right?
Mr Duncan: If two courses were fit for purpose, and one cost more matters,
   I would take the cheapest one, yes.
Mr Howe QC: Can I ask you not to get hung up on just the fit-for-purpose
   proposition, but if one takes a helicopter view of what AWT were raising, they were expressing profound concerns at the approach you
   were taking in balancing time and cost savings on the one hand and
   adherence to what they considered to be appropriate engineering
   standards on the other hand; do you agree that that is generally what
   they are doing?
Mr Duncan: Yes, but—
Mr Howe QC: I just want to know whether, in the light of the disagreement
   that arose between PTT on the one hand and AWT on the other hand,
   you basically continued with the same approach or whether the disagreement caused you to significantly revise your general approach?
Mr Duncan: I don't agree with most of the position in this document.
Mr Howe QC: Yes, so do we take it from that that the concerns that AWT
   was raising at the time you didn't think were well founded, and you
   pretty much continued to approach the balancing of those competing
   factors in the same way as you had done?
Mr Duncan: Yes, but the fit-for-purpose part is quite important in that.
   (Duncan 2010: 1390–1391)

Beyond not agreeing with Advanced World Technologies' position,
there was very little evidence in the hearing of what Mr Duncan did in
response to their accusations. It does not, at least on the face of the evidence, seem to have raised any *"red flags"*, nor caused him to make any
further enquiries as to whether or not PTT may have in fact been drifting
to a position where it was putting production before safety.

Deepwater Horizon

Similar issues have emerged in the various inquiries arising out of the Deepwater Horizon event; however, one example can be used to illustrate the case.

On 14 June 2010, the Committee on Energy and Commerce, a committee of the House of Representatives on the United States Congress, wrote to the then chief executive officer of BP, Tony Hayward (Waxman and Stupak 2010). The letter described a number of areas of concern that the Committee was seeking information about as part of its investigation into BP's role in the Deepwater Horizon disaster. One area was the use of centralizers used to position the drill string in the well bore:

> Centralizers. When the final string of casing was installed, one key challenge was making sure the casing ran down the center of the well bore. As the American Petroleum Institute's recommended practices explain, if the casing is not centered, "it is difficult, if not impossible, to displace mud effectively from the narrow side of the annulus", resulting in a failed cement job. Halliburton, the contractor hired by BP to cement the well, warned BP that the well could have a "SEVERE gas flow problem" if BP lowered the final string of casing with only six centralizers instead of the 21 recommended by Halliburton. BP rejected Halliburton's advice to use additional centralizers. In an e-mail on April 16, a BP official involved in the decision explained: "it will take 10 hours to install them. ... I do not like this." Later that day, another official recognized the risks of proceeding with insufficient centralizers but commented: "Who cares, it's done, end of story, will probably be fine." (Waxman and Stupak 2010: 2)

The BP official who commented *"it will take 10 hours to install them. ... I do not like this"* was Alexander Guide, a BP wells team leader. Mr Guide gave evidence to the Joint Investigation (Guide 2010) into the Deepwater Horizon incident on 22 June 2010, and a number of questions focused on the decision to only use six centralizers instead of the 21 recommended by Halliburton.

Mr Mathews: Do you see the e-mail from John Guide to Gregory Walz?
Mr Guide: Yes, sir.

Mr Mathews: In the middle of the page. Can you read that for me please?

Mr Guide: It says "I just found out the stop collars are not part of the cen-tralizers as you stated. Also it will take ten hours to install them. We are adding 45 pieces that can come off as a last minute addition. I do not like this and, as David approved in my absence, I did not question it. But now I am very concerned about using them."

Mr Mathews: Okay. What was the concern about using them?

Mr Guide: We had a similar -- I'm sorry, one of the other rigs in our fleet it was in the Atlantis Field several days before this had run a string of casing or attempted to run a string of casing and they had issues. And they had to pull the casing back out of the hole. And, when they pulled the casing back out of the hole, they left numerous centralizers in the hole.

Mr Mathews: Okay.

Mr Guide: The centralizers that are sent out usually and on this case the regular centralizers that are in the plan are actually centralizer subs that are made up onto the casing. Because of the incident that we had on the Atlantis, well I was, you know, concerned about -- since we didn't know exactly what happened, a reoccurring issue.

Mr Mathews: Okay. So the concern of having ten hours to install them was not the final decision on why that was not run?

Mr Guide: No, sir.

Mr Mathews: Was there a reason why you indicated ten hours to install them?

Mr Guide: It just was a reference to -- I didn't think it was prudent to take ten hours to install the wrong pieces of equipment. (Guide 2010: 67–68)

...

Mr Godwin: Had BP wanted to perform the complete installation of that casing in a manner that would have been safe, the job could have been shut down and waited for so many hours or a day or whatever to have gotten centralizers out there that it preferred, could it not? Could it not, sir? Yes or no?

[Legal objections]

Mr Godwin: Sir, isn't it true that if BP on the 16th of April decided that the 15 Weatherford centralizers were not the right ones, if it wanted to put safety first, it could have stopped the job, shut it down and said "We're going to get them in. We ordered those out here. We're going to make sure now we get the right ones." That could have happened, couldn't it?

[Legal objections]

Mr Godwin: Could BP have shut the job down when it realized it needed -- it wanted to have what it believed to be the correct centralizers?

[Legal objections]

Mr Godwin: I now want to know if at that time could they have shut the job down because the casing had not been completed.

[Legal objection and direction to answer the question]

Mr Guide: Okay. There was -- yes, BP could have shut the job down, but in no way, shape or form was safety ever compromised or -- or any part of any decision of compromising safety in this operation. (Guide 2010: 378–379)

From a personal management perspective, the point is not whether the decisions were right or wrong, or contributed to any unsafe conditions that may have arisen. Rather, how do you demonstrate that the decisions you made were well founded, risk based, and justifiable, because rightly or wrongly an assertion that *"safety was not compromised"* in the wake of a major accident event like the Deepwater Horizon sounds at best misguided, and at worst somewhat contrived.

Perhaps the best that could be said is that safety was never knowingly compromised.

Conclusion

This chapter has described a number of discrete examples of failures by managers to adequately respond to warning signs of possible failures in the health and management system of their organisations.

These types of warning signs present an opportunity to stop and consider the risks in the business. During the Montara Inquiry, Counsel assisting the inquiry made this point directly. Having identified a number of warning signs of problems in the operations being conducted, Counsel assisting put the following:

Mr Howe QC: That is a pretty obvious orange light or signpost for the carrying out of some considered review or audit; would you agree?

Mr Jacob: Certainly the three of them taken together, yes.

Mr Howe QC: The three of them taken together mount, do they not, an absolutely incontrovertible case for PTT pausing, in a timely fashion, and carrying out a considered review as to what had actually taken place out on Montara with respect to well control?

Mr Jacob: Yes, you would like to think so. (Jacob 2010: 1371–1372)

A number of the cases articulate what is, in essence, the minimum expectation of managers when confronted with a warning sign—that they will give it due and careful attention:

> A failure of float valves is a significant prob-
> lem which requires a thoughtful and considered
> response to the particular circumstances surround-
> ing that failure. (Borthwick 2010: 57)

And

> It was a serious memorandum which merited seri-
> ous thought and attention, and called for a consid-
> ered reply. (Department of Transport 1987: 8)

Clearly, when it comes to safety, warning signs are serious matters and they do warrant serious attention.

References

Borthwick, D. 2010. *The report of the Montara Commission of Inquiry.* Montara Commission of Inquiry, Canberra. http://www.ret.gov.au/Department/Documents/MIR/Montara-Report.pdf (accessed 25 November 2010).

CSB (U.S. Chemical Safety and Hazard Investigation Board). 2007. *Investigation report: Refinery explosion and fire.* Washington. http://www.csb.gov/assets/document/CSBFinalReportBP.pdf (accessed 23 November 2010).

Department of Transport. 1987. *MV Herald of Free Enterprise report of Court No. 8074 Formal Investigation.* London. http://www.maib.gov.uk/publications/investigation_reports/herald_of_free_enterprise/herald_of_free_enterprise_report.cfm (accessed 23 November 2010).

Duncan, C. 2010. *Transcript: Montara Commission of Inquiry.* http://www.montarainquiry.gov.au/transcripts.html (accessed 29 September 2010).

Guide, J. 2010. *Transcript: Deepwater Horizon Joint Investigation,* 22 July 2010: http://www.deepwaterinvestigation.com/go/doctype/3043/56779/.

Jacob, A. 2010. *Transcript: Montara Commission of Inquiry.* http://www.montarainquiry.gov.au/transcripts.html (accessed 29 September 2010).

McInerney, P. 2003. *Special commission of inquiry into the Waterfall Rail Inquiry.* Ministry of Transport. NSW. http://www.transport.nsw.gov.au/inquiries/waterfall.html (accessed 21 October 2010).

Rogers et al. 1986. *Report of the Presidential Commission on the Space Shuttle Challenger Accident.* http://history.nasa.gov/rogersrep/v1ch6.htm (accessed 23 November 2010).

Waxman, A., and B. Stupak. 2010. Letter from the U.S. House of Representatives subcommittee on Oversight and Investigations to Tony Hayward, Chief Executive Officer BP PLC dated 14 June 2010. http://energycommerce.house.gov/documents/20100614/Hayward.BP.2010.6.14.pdf (accessed 23 November 2010).

chapter 9

Learning lessons

> "The Panel considers the similarities between the 'lessons' from Grangemouth and the Texas City incident to be striking ..."[*]

Introduction

Learning lessons are related to but different from warning signs. Warning signs are a direct indicator of something happening in your organisation right now, in the case of Montara the letter from contractors accusing PTTEPPAA of putting production before safety, for example.

Identifying and learning lessons is more about looking at incidents that occur in different parts of a business, or indeed in other businesses or other industries, and working out what we can learn from those incidents and how the lessons might apply to our business and the parts of the business that we are responsible for.

In this context lessons can be less direct. Learning lessons may take more effort, as it requires looking at an event and determining what application, if any, it has to your organisation.

In many ways, the discussion in this book, and the common failings identified in major accidents regardless of geography, time, or industry, adds to an already compelling body of evidence that organisations find it very difficult to learn safety lessons, or if they are learned, very difficult to implement and maintain the lessons.

In this chapter I will look at examples where organisations had experienced a serious incident, but where the lessons of that incident were not learned, or not embedded and the failures repeated—often with significantly worse consequences.

BP Texas City

Five years before the fire and explosion at Texas City, there had been another serious incident at one of BP's refineries, in Grangemouth, Scotland. The incident involved a large fire and a number of serious process upsets.

---

[*] Baker et al. 2007: 183.

The UK Health and Safety Executive investigated the incident and released a major investigation report in August 2003 (HSE 2003), and many of the key findings in that investigation re-emerged two years later at Texas City. Significant findings in the Grangemouth Investigation included:

- Management inattention and a lack of resources directed to *"maintaining and improving technical standards for process operations and enforcing adherence to standards, codes of practice, good engineering practice, company procedures and the HSE guidance"*
- Inadequate maintenance of the integrity of high hazard plant
- No effective process safety review
- Lessons from major incidents not adequately actioned (HSE 2003: 65)

The Grangemouth Investigation identified that "Lesson 1" of the key lessons for *Major Accident Hazard Sites* arising from the investigation was:

> Major accident hazards should be actively managed to allow control and reduction of risks. Control of major accident hazards requires a specific focus on process safety management over and above conventional safety management. (HSE 2003: 74)

However, it seems that these lessons did not make their way to Texas City, even though an internal investigation conducted by BP identified failures around the lack of specific focus on process safety management and the impact of cost, and then published these findings internally within BP (CSB 2007: 144–145).

Indeed:

> The CSB found that a number of managers, including executive leadership, **had little awareness or understanding of the lessons from Grangemouth**. Moreover, BP Group leadership did not ensure that necessary changes were made to BP's approach to safety. They did not effectively address the need for greater focus on PSM, including measuring PSM performance, nor did they resolve problems associated with BP's decentralized approach to safety. (CSB 2007: 144–145) [emphasis added]

The Baker Panel Review also reinforced the lack of learning, noting that BP's response to Grangemouth was an opportunity missed (Baker et al. 2007: 2004).

Piper Alpha

In September 1987 a fatality occurred on Piper Alpha that contained potential lessons that pointed to weaknesses in the safety management system that eventually played out in the disaster less than 12 months later.

In what the Piper Alpha Inquiry referred to as the "Sutherland fatality" (Cullen 1990: 197), Mr Sutherland, a rigger employed by a contractor, was killed. Lord Cullen stated that:

> The accident and what arose out of it has a significant bearing on the discussion of the adequacy of [the] attention to the quality of [the] permit to work system and handover procedure. (Cullen 1990: 197)

The work being performed at the time of the accident required a motor to be lifted to repair a damaged bearing. A system for the lift was devised during the day shift with the assistance of qualified riggers. The night-shift mechanical technicians decided to change the method of lifting without discussing the change with the night-shift lead—the expert who could have advised on the new system.

Mr Sutherland climbed on to a panel to attach a shackle for the lift. The panel shifted, and Mr Sutherland fell and was injured. He later died.

The permit that had been issued for doing the work stated simply *"check and repair the … bearing"*. An internal inquiry into the event identified that *"the expansion of the original scope of the work to the extent that it required the raising of the motor did not alert the supervisor to the additional measures that might have been taken to ensure the safe conduct of the new work scope"* (Cullen 1990: 197–198).

The company was charged in relation to the fatality, with one of the particulars of the charge being that no new permit was taken out to cover the installation of the said lifting gear and other necessary work. Lord Cullen formed the view that once again it seemed that the approach left too much to be settled as the work went along.

Another particular of the charge against the company was that they did not have adequate communication of information between the day and night shifts. Although the Company pleaded guilty to this charge, evidence in the Piper Alpha inquiry showed that no changes were made to the handover practices after the Sutherland and that there was *"no awareness of any weaknesses or criticism of the communication processes"* (Cullen 1990: 198).

There were other elements identified from the Sutherland fatality and even memoranda issued to address some of these deficiencies, but ultimately Lord Cullen found that, although action was taken in some respects after the fatality, this did not have a lasting effect in practice.

Lord Cullen was also critical of the internal inquiry into the incident, noting:

> It contains no examination of the adequacy or quality of the handover between the maintenance lead hands. Nor did it examine the implications of the expansion of the scope of work beyond what had been covered by the PGW. However, the [charges] to which [the company] had pleaded guilty specified that there had been a failure of supervision in both these areas. In my view the work with the board of inquiry was superficial in respect that it did not examine either of these areas, at the latest after [the company] had pleaded guilty ... if there had been an adequate handover or if the work had been limited to the scope and conditions of one or more permits to work the fatality could have been prevented. (Cullen 1990: 231)

Lord Cullen was also critical of the circulation of the findings from the Sutherland fatality, noting that safety personnel, including the safety supervisor on Piper Alpha, did not know what the management team had decided to do about the deficiencies identified from the fatality (Cullen 1990: 232).

Mr Bodie, the onshore safety superintendent, was asked whether he had made representations against a policy discouraging the discussion of accidents such as the Sutherland fatality. He replied:

> "We certainly had discussions. It really is a problem, having found out what had happened in any particular incident, then to have to disguise your writing and send out memos without any mention of that particular incident but try to get action taken."

> When asked whether that militated against the proper feedback which ought to have arisen he replied:

> "No, we managed to get the messages across to the personnel, in a lot of cases verbally, and, as I said, by very cleverly worded memos." (Cullen 1990: 233)

In the end, the Piper Alpha Inquiry was left unimpressed by the actions taken to share the lessons from the Sutherland fatality:

> In the result I considered that whether by direction
> or by inaction ... management failed to use the cir-
> cumstances of particular incidents to drive home
> the lessons of the investigation of those incidents to
> those who were immediately responsible for safety
> on Piper on a day to day basis. (Cullen 1990: 233–234)

Deepwater Horizon—Transocean

In December 2009 an incident occurred on one of Transocean's rigs operating in the North Sea, which the National Commission described as an
"eerily similar near-miss" (Graham et al. 2011: 124).

On that occasion also, gas entered the riser during operations, after
a negative-pressure test had been run and declared a success. Mud
spilled out onto the rig, but the well was able to be shut in before a
blowout occurred.

Transocean investigated the incident and circulated information
about its using an internal presentation. Some of the lessons from the 2009
incident identified in that presentation resonated strongly with findings
that emerged from the disaster in 2010. Findings such as

- tested barriers can fail; and
- negative pressure tests were not adequately covered in the well control documentation.

The presentation concluded by asking, *"Are we ready?"* and *"WHAT
IF?"* (Graham et al. 2011: 124).

Transocean also sent out an "operations advisory", setting out the lessons learned and warnings from the presentation.

According to the National Commission, neither the presentation nor
the advisory were provided to the Deepwater Horizon crew.

Despite Transocean's view that the 2009 incident and what occurred
in the Gulf of Mexico were different, the National Commission concluded
that they were linked:

> The basic facts of both incidents are the same. Had
> the rig crew been adequately informed of the prior
> event and trained on its lessons, events at Macondo
> may have unfolded very differently. (Graham et al.
> 2011: 125)

Conclusion

The irony in the case of the Transocean is that when the 2010 event was unfolding in the Gulf of Mexico, senior managers of BP and Transocean were on the Deepwater Horizon sharing safety lessons from other facilities:

Q: Sir, if I recall your earlier testimony, you indicated that part of your trip out to the Deepwater Horizon on this occasion was to conduct safety audits; is that correct?

A: That -- that is correct. If it presents itself that's one of the -- one of the things that -- that we would do.

Q: On this particular occasion did you conduct any safety audits and, if so, what type?

A: We -- we went to an area aft -- on the aft deck the rig riser would be laid down, if a riser was not in operation, not in use. Discussed an incident that has occurred on another rig and talked about how if that same situation could happen on the Horizon and if -- you know, and what the differences were, if they had learned or adopted any learnings from that incident. We went up to look at the thing called the skate that moves pipe from that area up to the rig floor and back and there's an area underneath it, kind of a bay underneath, that was the scene of an incident that had happened on another rig just a few days -- a few days or a week before our visit. And we went to that area to see, again, how the Horizon would -- would be different or similar to this other rig. And, again, to test whether that learnings from that incident had been communicated and how the Horizon had dealt with those, with those learnings.

Q: Sir, you mentioned that one of the areas you visited was the skate and a prior incident. Could you give us a little more detail about that particular incident and what the lessons learned were?

A: Well, the incident was -- it was a hydraulic line that had -- was leaking in that -- in that area on the other rig. And they -- one of the drilling contractor personnel had **stepped down into that area with the charter material to clean up the leak and lost his footing catching himself he had -- I believe he hurt his shoulder.** The learnings from that were around -- as we almost always find learnings are around hazard recognition. Making sure that you think about where you're going to step or where you're going to put your hand. We recognize that access to that area was -- could be improved. And, again, these were investigations that are led by Transocean and BP participates sometimes in these. What we do is we both embrace the learnings and try to make sure those learnings get passed from rig to rig. But those were a couple of the learnings. We went to the Horizon to look and see if those same -- if the same learnings about applying non-slip material would -- would apply, whether access to the area was the same. In fact on the Horizon it was a little bit deeper skate. It was a little bit more difficult to get into. We discussed what Transocean was going to do relative to -- to the event, the incident. (Sims 2010: 185–186) [emphasis added]

There is a further irony in that at the time senior management were concerned to pass on lessons about slip, trip and fall hazards, they also had an opportunity to engage with workers about difficulties they were having conducting pressure tests—tests that proved to be critical to the eventual disaster (Hopkins 2011).

It is tragic that in focussing on personal safety risks—slips, trips and falls—the opportunity to avert a disaster may have been lost. Doubly so, when you consider that one of the most important lessons to come out of the Texas City disaster was the need for managers to lift their gaze above slips, trips and falls and to look at the critical risks in a business.

References

Baker, J. et al. 2007. *The report of the BP U.S. Refineries Independent Safety Review Panel*. U.S. Chemical Safety and Hazard Investigation Board, Washington. http://www.bp.com/liveassets/bp_internet/globalbp/globalbp_uk_english/reports_and_publications/presentations/STAGING/local_assets/pdf/Baker_panel_report.pdf.

CSB (U.S. Chemical Safety and Hazard Investigation Board). 2007. *Investigation report: Refinery explosion and fire*. Washington. http://www.csb.gov/assets/document/CSBFinalReportBP.pdf (accessed 23 November 2010).

Cullen, Lord. 1990. *The public inquiry into the Piper Alpha disaster*. Department of Energy. London: HMSO.

Graham, B. et al. 2011. *Deep water: The Gulf oil disaster and the future of offshore drilling*. Report to the President. National Commission on the BP Deepwater Horizon Oil Spill and Offshore Drilling. http://www.oilspillcommission.gov/final-report (accessed 11 January 2011).

Hopkins, A. 2011. *Management walk-arounds: Lessons from the Gulf of Mexico oil well blow out*. Working paper 79. National Research Centre for OHS Regulation. Australian National University. Canberra. http://ohs.anu.edu.au/publications/pdf/WP%2079%20Hopkins%20Gulf%20of%20Mexico.pdf (accessed 18 March 2011).

HSE (Health and Safety Executive). 2003. *Major incident investigation report, BP Grangemouth, Scotland, UK*. http://www.hse.gov.uk/comah/bpgrange/contents.htm (accessed 21 July 2010).

Sims, D. 2010. *Transcript: Deepwater Horizon Joint Investigation*, 29 May 2010. http://www.deepwaterinvestigation.com/go/doctype/3043/56779/.

chapter 10

Managing change

"After the March 2005 incident, BP determined
that a majority of the mobile officer trailers were
sited without applying the [management of change]
process"[*]

Introduction

If there is one constant in modern workplaces, it is change—a lot of it
and the potential for it to occur very quickly. When change happens, the
systems that have been put in place to manage workplace safety risks may
very quickly become obsolete. Indeed, they may increase the risk of a
workplace accident.

Modern research into workplace accidents has identified that missing
the side effects of change is the most common form of failure for individuals and organisations (Hollnagel, Woods and Leveson 2008).

The Herald of Free Enterprise

The influence of change was clearly evident in the *Herald of Free Enterprise*
case, where a range of changes were made to accommodate differences
in operations between the ports at Calais and Zeebrugge, all of which,
cumulatively, combined to place pressure on the individuals carrying out
their duties and contributed to the accident:

> Thus at Zeebrugge the turn-round was different
> from the turn-round at Calais in four main respects.
> At Zeebrugge (1) only two deck officers were avail-
> able, (2) only one deck could be loaded at a time,
> (3) it was frequently necessary to trim the ship by
> the head, and (4) the bow doors could be closed at
> the berth. Because of these differences, with proper
> thought the duties of the deck officers at Zeebrugge
> would have been organized differently from their
> duties at Calais. No such thought was given to the

---

[*] CSB 2007: 127

> matter, with the result that immediately loading
> was complete the Chief Officer felt under pressure
> to leave G deck to go to his harbour station on the
> bridge. (Department of Transport 1987: [17.4])

As changes occur within an organisation, they can put increasing pressure on the resources of that organisation, causing it to move into a state where accidents are more likely to occur. Over time, these changes may accumulate slowly, making it difficult for the organisation to detect that it is moving to a "less safe" way of operating. Because of this, managers need to always be alert, mindful of warning signs in the business that health and safety risks are not being managed as the organisation expects.

Managing change presents a number of difficult and substantial challenges for managers.

Although there is a need to develop effective processes to identify the need for and then implement good change management practices, there is also a need to recognise their limitations. As the various examples in this chapter illustrate, the potential impact of change on the safety of an operation can arise at all levels within the business, from governance structures to supervisors leaving early, and even, in the case of government approaches to privatization and de-regulation, externally. As we have previously discussed in the context of rules more generally, it is unlikely that an organisation would ever develop policies and procedures that could identify every possible combination of risk and circumstances which might give rise to an accident, and even if it could, no one could possible read, remember, or regularly refer to it. This limitation means that managers themselves have to have—and then generate within their workforce—an awareness of change and a preparedness to question and challenge change in the interests of ensuring that it is safe.

As the Longford Royal Commission illustrates, even when high quality documented systems are in place, people still need the knowledge and awareness to implement them.

Change can impact an organisation in so many different ways and on so many different levels that the entire organisation needs to be mindful of the possible threats posed by change.

At one level, change can be immediate with a close link between cause and effect, such as the decision to remove and not replace the 9⅝" PCCC from the H1 well in the Montara well blowout. But in other cases, change may be quite distant in time and/or geography from an incident. Examples of such change might include changing the roles and responsibilities of supervisors or removing personnel from site, but in all cases managers need to give proper thought and due attention to all changes in the organisation to try to understand the actual or potential impact that those changes might have on the safe operations of the organisation.

We have already touched on an element of managing change in the context of rules: normalisation. Normalisation represents a gradual drift over time in the way that things are done, as illustrated in the Blackhawk 221 case—a pushing of the envelope so that things are eventually being done in a way that was not anticipated by the system, which does not manage the new risks created by the change.

Normalisation is a different type of change to manage, principally because it does develop over time and there is no clear change point that triggers (or potentially triggers) a reconsideration of the risks and the way that they are managed. In the analysis and monitoring of safety management, one needs to be alert for indicators of change, and when detected, change should be seen as an opportunity to re-examine the critical elements of the safety management process. Key questions include:

1. Does the change impact existing hazards and/or create new hazards?
2. Does the change alter any of the risks arising from existing hazards, and does it create any new risks?
3. Are the controls that are in place to manage the risks still suitable and effective in light of the change?

It may also be necessary to consider whether or not the change is acceptable regarding the hazards and risks affected by or arising from the change.

The impact of change has manifested itself in any number of ways across a broad range of industries. Changes that have formed part of the causal mix in workplace accidents include:

- Staff reductions or relocations
- Budget cuts
- Process changes that did not incorporate changes to equipment, with that equipment then being rendered unsuitable
- Changes in personnel roles and responsibilities
- Changes in equipment to be used at the workplace
- Changes in corporate organisation and governance structures

These types of changes could generally be managed as middle- to upper-level management decisions, perhaps even "macro" changes, but there are also numerous examples of "micro" or workplace level changes that need to be accounted for and managed. Examples include:

- Supervisors having to leave work early for a family emergency and not being replaced
- Changes to work schedules and tasking

Longford

The Longford gas plant explosion provides a number of examples of the type of changes and management decisions that can be subject to criticism if they are not thoroughly considered and risk-assessed for their potential impact on safety. During the investigation into the explosion, the Royal Commission identified four changes which should have been subject to a management of change risk assessment but were not. These included changes to the plant design in late 1992, a decision to relocate engineers from the Longford plant back to Melbourne, changing the roles and responsibilities of supervisors at Longford, and reducing the numbers of maintenance personnel (Dawson and Brooks 1999: 210).

The systems in place at Longford recognised the likelihood of and need for change within their operations, while at the same time accepting that *"changes potentially invalidate prior risk assessments and can create new risks, if not managed diligently"* (Dawson and Brooks 1999: 206) and processes were in place to assure that risk assessments were conducted as a part of any change process. The Royal Commission did make some criticism of the processes and procedures in place for the management of change, but they are not especially important in the context of this discussion, given that the management of change processes was not applied at all.

Montara and the Deepwater Horizon

Both the Montara Inquiry and the National Commission of Inquiry into the Deepwater Horizon reached similar conclusions about change in the context of those disasters: in both cases the emphasis was on the apparent culture of production over safety, and in both cases it was found that management made decisions that increased safety risk without giving adequate or proper consideration to possible safety risks arising from the decisions that were made. The role of production over safety is discussed in more detail later, but as the National Commission observed:

> There is nothing inherently wrong with choosing a less-costly or less-time-consuming alternative—as long as it is proven to be equally safe. The problem is that, at least in regard to BP's Macondo team, there appears to have been no formal system for ensuring that alternative procedures were in fact equally safe. (Graham et al. 2011: 125)

Both the Montara and the Deepwater Horizon disasters were infected by a seeming inattention to the risks arising from changes.

The Montara Commission of Inquiry identified that PTTEPAA had a prevailing philosophy of getting the job done without delay and repeatedly made decisions where risks were not recognised when they should have been and not assessed properly when recognised (Borthwick 2010: 11).

The National Commission of Inquiry into the Deepwater Horizon disaster set out a table of decisions that the Inquiry found were likely to have increased risk at the Macondo well whilst at the same time potentially saving time (meaning money) and that were not subject to comprehensive and systematic risk-analysis, peer-review, or management of change processes (Graham et al. 2011: 125).

Montara

In the Montara Inquiry one of the key change issues, if not *the* key change issue, arose from the use and removal of PCCCs.

When suspending an oil or gas well, it is important that adequate barriers be put in place to stop the hydrocarbons entering the well and making their way uncontrolled to the surface. In the case of Montara, the original well suspension plan relied on the cemented casing shoe at the bottom of the well and two additional cement plugs variously positioned at points in the well. The well suspension plan was changed, and a decision was made to use another type of barrier: pressure-containing anti-corrosion caps (PCCCs).

Ultimately, the Montara Inquiry determined that the changes leading to the use and removal of the PCCCs arose because of production and cost pressures—production before safety. (We will later look at the way this specific issue was considered at that time, in Chapter 11, "Production before Safety".)

Another critical issue involving the PCCCs was a decision to remove one of the PCCCs, the 9⅝" PCCC.

The change to the well suspension plan had called for the installation of two PCCCs, the 9⅝" PCCC and the 13⅜" PCCC. For reasons that were never made clear during the Montara Inquiry, the 13⅜" PCCC was never installed. When this was discovered, a decision was made to remove the 9⅝" PCCC from the well so that some of the casing in the well could be cleaned.

The removal of the 9⅝" PCCC was a change in the plan process; the 9⅝" PCCC was not due to be removed until 24 August 2009 as part of the originally planned process. As a result, the Commission found:

> If all of those events had taken place as planned, it
> is likely that the deficiencies in the cemented cas-
> ing shoe would have been detected as a result of
> the forecast pressure test. At that point, the derrick

would have been over the H1 Well and therefore urgent remedial steps could have been taken to prevent a blowout. Mr Gouldin emphasised the significance of this in his evidence.

As it happened, the removal of the 9-5/8 PCCC was brought forward by four days and four hours, and that removal was not followed by any testing of the 9-5/8 casing stream. This had the effect of leaving the H1 Well wholly dependent upon the cemented casing shoe as the only barrier against a blowout—a barrier which had not been properly tested and verified. (Borthwick 2010: 103)

The nature of this change was ultimately accepted by PTT as being a significant one:

Mr Howe QC: So far as you're aware, a proper risk assessment would have entailed a decision to reinstall the 9-5/8" [PCCC] at the earliest practicable opportunity; that's right, isn't it?

Mr Jacob: I would have expected that, yes.

Mr Howe QC: There can be no sensible justification for doing otherwise, can there?

Mr Jacob: Not that I'm aware of at the moment.

Mr Howe QC: Do you recall the evidence that to reinstall the 9-5/8" [PCCC] would have occupied as little as 15 minutes, perhaps half an hour?

...

Mr Jacob: Yes.

Mr Howe QC: So does it seem extraordinary to you, looking back on the events which occurred on 20 August that the 9-5/8" [PCCC] was left off the H1 well?

Mr Jacob: Yes. (Jacob 2010: 1660–1661)

Unsurprisingly, the decision to remove and not replace the 9⅝" PCCC was also subject to some scrutiny by the Inquiry.

Mr Howe QC: Before taking that decision to remove the 9-5/8" PCC, did you undertake a careful risk assessment of the potential implications of so doing?

Mr Duncan: I undertook a brief risk assessment, yes, but it was not adequate in the light of what happened. (Duncan 2010: 1291)

...

Mr Howe QC: And you did not undertake a single inquiry with respect to that topic before you removed the 9-5/8" cap; that's right, isn't it?

Mr Duncan: That's correct. (Duncan 2010: 1292–1293)

...

Mr Howe QC: Had you undertaken a proper risk assessment, you might have ascertained that in fact the 9-5/8" PCC hadn't been tested or verified post installation; that's right, isn't it?

Mr Duncan: That's possible.

Mr Howe QC: So all the manner of care in terms of its removal, about checking for pressure underneath, might have amounted to nothing, because it could have been damaged before installation or in the course of installation; that's right, isn't it?

Mr Duncan: That's true.

Mr Howe QC: And you hadn't excluded that possibility?

Mr Duncan: No, I hadn't excluded the possibility, no. (Duncan 2010: 1293–1294)

Complexity of change

Regarding these questions and answers, it would be easy to write off the decision to remove and not replace the 9⅝" PCCC as a wholly incompetent, irresponsible, and entirely production-focused decision that had no reasonable or meaningful regard to safety.

To do so would be dangerous and risk sidelining the valuable lessons from these decisions as merely being aberrations.

Mr Duncan, the manager who made the decision to remove the 9⅝" PCCC, was an experienced and well-regarded operator in the oil and gas industry, yet in the face of the disaster and the role that the removal of the 9⅝" PCCC played, he was (at least initially) adamant that his decision, based on the information that he had at the time, was reasonable.

Mr Howe QC: Before taking that decision to remove the 9-5/8" PCC, did you undertake a careful risk assessment of the potential implications of so doing?

Mr Duncan: I undertook a brief risk assessment, yes, but it was not adequate in the light of what happened.

Mr Howe QC: No. Well, I suggest to you that it wasn't adequate in the circumstances that obtained at the time; do you agree with that?

Mr Duncan: No.

...

Mr Howe QC: You think you acted perfectly reasonably in the circumstances, do you?

Mr Duncan: In the circumstances, the missing 13-3/8″ corrosion cap was not good, but the removal of the 9-5/8″ cap was unrelated to that. (Duncan 2010: 1291–1292)

...

Mr Howe QC: You found out that one secondary barrier was in place, and you decided to remove the only other secondary barrier that was in place; that's right, isn't it?

Mr Duncan: That's correct.

Mr Howe QC: Which made the entire control of the well dependent upon the cemented casing shoe; that's right, isn't it?

Mr Duncan: That's correct.

Mr Howe QC: And you did not undertake a single inquiry with respect to that topic before you removed the 9-5/8″ cap; that's right, isn't it?

Mr Duncan: That's correct.

Mr Howe QC: I want to suggest to you, sir, that in the circumstances that obtained at the time, you should have undertaken such an inquiry; do you accept that?

Mr Duncan: No, I don't. (Duncan 2010: 1292–1293)

...

Mr Howe QC: And do you accept that, to that extent, your approach to the matter was seriously deficient?

Mr Duncan: Yes, but I'll explain, when you want. (Duncan 2010: 1295)

Deepwater Horizon

We should also consider the cross examination of Mr Guide in relation to his decision to proceed with only six centralisers rather than the 21 recommended by Halliburton.

Mr Godwin: Could BP have shut the job down when it realized it needed -- it wanted to have what it believed to be the correct centralizers?

...

Mr Godwin: I now want to know if at that time could they have shut the job down because the casing had not been completed.

...

Mr Guide: Okay. There was -- yes, BP could have shut the job down, but in no way, shape or form was safety ever compromised or -- or any part of any decision of compromising safety in this operation. (Guide 2010: 378–379)

Bad decisions?

Leaving aside the difficulties that both men must have faced in terms of the general public scrutiny they were under and the specific pressure of being subject to intense and very public cross-examination, personal criticism, and the risk of personal liability, it may not have been unsurprising that both men found it difficult to accept that their decisions (at the time they were made) compromised safety. In fact, from a safety management perspective, it would be far more helpful to start from the position that both Mr Duncan and Mr Guide were entirely competent and experienced operators who made what they believed to be entirely reasonable decisions, and then try to understand how what they believed at the time to be reasonable and safe had such adverse outcomes.

I think, for both men, there can be little doubt that the production (time and cost) pressures played a role in both decisions. This is discussed in more detail later, but for now I am not suggesting in any way that either man made a conscious decision to choose a cheaper but less safe option. There is nothing that I have seen in any of the evidence or in anything presented in any of the inquiries that could suggest any such deliberate recklessness by either man. Mr Duncan, in particular, was on the rig at the time he made the decision to remove the 9⅝" PCCC, and I could only assume that he would have had a very vested interest in ensuring that things were done as safely as possible.

So, let's leave to one side for the moment the issue of production over safety.

I think that the critical failure, both in this case and in the case of most safety management failures, is the failure to appreciate the complexity of safety management, and without an appreciation of complexity it is easy to understand why, for example, Mr Guide would be adamant that he did not make any decisions that compromised safety.

At the National Commission of Inquiry into the Deepwater Horizon disaster identified, *"complex systems almost always fail in complex ways"* (Graham et al. 2011: viii). This was true of the *Columbia* Space Shuttle accident referred to by the National Commission of Inquiry, and it is also true of almost every major accident event.

If there had been no other failures of the safety management system on the West Atlas or the Deepwater Horizon, then the decision by Mr Guide to only use six centralisers and the decision by Mr Duncan to remove and not reinstall the 9⅝" PCCC would both have been unlikely to lead to anything further. Indeed, in Mr Duncan's case if the safety management system had worked as expected, the 13⅜" PCCC would have been in place and there would have been no reason for the 9⅝" PCCC to have been removed.

At the time, and in isolation, the decisions quite likely were not seen to compromise safety. If neither man was *aware* (or should have been aware) of all of the combinations of circumstances surrounding their decisions (and that may be debatable), is it fair to point at those decisions as indicating that they were incompetent or reckless? I think that it is at this point that we can start to get a very real sense of the complexity surrounding management of change brought about by two competing perspectives— one of which was dealt with in the two inquiries, one of which was not— and I think that it is the one that was not dealt with that is more important.

Both inquiries were, on any fair reading of the evidence, able to establish that the decisions that were made were objectively likely to increase risk. In the case of the decision to remove the 9⅝" PCCC, the H1 well was left with only one barrier between the reservoir and the top of the well. This must be seen as an objectively less safe option than the two barriers that would have been in place if it had been left in. In the case of the number of centralisers, again, it seems objectively clear in the face of a warning that the well would have a "severe gas flow problem" if fewer than the 21 recommended centralisers were used, and in the absence of any meaningful evidence to show that using six was safe, using six must have been objectively riskier.

But if these decisions were "bad" decisions, how many times had they or other managers made similar decisions that were "good" decisions? Decisions that got the job done? Moved the organisation forward? Enabled them to rise to the levels that they had within the industry that they worked in?

To what extent were they working normally? And to what extent was there anything that would have put them on notice that the way they made decisions generally was not acceptable (if indeed they were not) or that they should have been doing something different in this case?

References

Borthwick, D. 2010. *The report of the Montara Commission of Inquiry.* Montara Commission of Inquiry, Canberra. http://www.ret.gov.au/Department/Documents/MIR/Montara-Report.pdf (accessed 25 November 2010).

Dawson, D., and B. Brooks. 1999. *Report of the Longford Royal Commission: The Esso Longford gas plant accident.* Melbourne: Government Printer for the State of Victoria.

Department of Transport. 1987. *MV Herald of Free Enterprise report of Court No. 8074 Formal Investigation.* London. http://www.maib.gov.uk/publications/investigation_reports/herald_of_free_enterprise/herald_of_free_enterprise_report.cfm (accessed 23 November 2010).

Duncan, C. 2010. *Transcript: Montara Commission of Inquiry.* http://www.montarainquiry.gov.au/transcripts.html (accessed 29 September 2010).

Guide, J. 2010. *Transcript: Deepwater Horizon Joint Investigation,* 22 July 2010. http://www.deepwaterinvestigation.com/go/doctype/3043/56779/.

Graham, B. et al. 2011. *Deep water: The Gulf oil disaster and the future of offshore drilling*. Report to the President. National Commission on the BP Deepwater Horizon Oil Spill and Offshore Drilling. http://www.oilspillcommission. gov/final-report (accessed 11 January 2011).

Hollnagel, E., D. Woods, and N. Leveson. 2008. *Resilience engineering concepts and precepts*. England: Ashgate.

Jacob, A. 2010. *Transcript: Montara Commission of Inquiry*. http://www.montarainquiry.gov.au/transcripts.html (accessed 29 September 2010).

chapter 11

Production before safety

> "Okay. There was -- yes, BP could have shut the job down, but in no way, shape or form was safety ever compromised or -- or any part of any decision of compromising safety in this operation."*

Introduction

One of the dominant themes that emerge in most major accident inquiries is "production before safety". Production before safety manifests itself in many different ways: from accepting the cheapest bidder, to shaving precious minutes off schedules and deadlines; from choosing less expensive equipment to accepting the way that work has been completed and not being prepared to spend the time to go back and check that it was completed safely.

A common difficulty, however, is that there is seldom a clear line of thinking between cost or time or production and safety; in other words, decision makers are not deliberately making (what they know to be) unsafe decisions in the light of a cost or production benefit. Rather, it is a combination of circumstances, including the culture of an organisation, the knowledge and competence of the decision maker, and any other numbers of factors, that result in decisions being made that erode the safety margins of an organisation.

For a manager, one of the challenges is to recognise where decisions could be made that might result in a tension between production and safety, and to have in place appropriate trip wires, or red flags, to recognise when those decisions are made—and once recognised, to have in place good processes to have assurance that safety is not being unreasonably compromised.

Businesses naturally operate at the edge of their resources—competition and markets drive the need for efficiency, and the drive for efficiency places stress on all aspects of an organisation, including safety. This is not inherently wrong, nor inherently unacceptable. The National Commission of Inquiry into the Deepwater Horizon disaster recognised this:

---

* Guide 2010: 378–379

There is nothing inherently wrong with choosing a less-costly or less-time-consuming alternative—as long as it is proven to be equally safe. (Graham et al. 2010: 125)

Risk assessment

One of the common characteristics of the way that "production before safety" is revealed in major accident investigations is the lack of meaningful risk assessments of decisions that require a balancing of production and safety.

Both the Deepwater Horizon and Montara inquiries were explicit in the examination of production and safety, pointing in each case to decisions that were made which saved the relevant companies money (and/or time, which on a drilling rig amounts to the same thing) but were found to have increased safety risks. Further, in both cases it was usually the case that these cost-saving decisions were not subject to adequate or proper risk assessments to fully understand the potential safety impacts.

Both of these inquiries, like so many others, were also characterised by the types of arguments and findings that arise in any discussion about "production before safety"—findings, for example, that there was an organisational culture of production before safety, rather than an explicit finding that any one individual *knowingly* chose a riskier, less expensive option, conscious that he or she was making an unsafe decision; and arguments by individual managers and decision makers that their decisions did not compromise safety and insistence by senior managers that putting production or costs before safety was not tolerated in their organisation.

That is not to say that cost savings and efficiency cannot exist without compromising safety: they can (as evidenced by the many organisations operating successfully and viably without compromising safety) and must in order for the business to progress. However, what is often missing in the safety/efficiency balancing act is assurance. How do managers at all levels of the organisations get assurance that their efficiency measures, their business planning, their incentive programs, and their reward and recognition strategies do not compromise safety?

A simple assertion by senior management that it *must not* is clearly insufficient.

What positive assurance processes are in place to provide the link between the aspirational desire to be cost effective without compromising safety and the actual practices being undertaken when cost and safety are overlapping factors in business decisions?

The Herald of Free Enterprise

We have already seen a number of instances in management failures leading to the *Herald of Free Enterprise* disaster, and unsurprisingly, production before safety, in the form of "on-time running", also featured as a causal factor in the accident.

There was clear evidence in the inquiry that the management of Townsend drove a sense of urgency in the organisation to ensure that the ferries got away from port as quickly as possible. The officers and men of the *Herald of Free Enterprise* were not immune from this pressure:

> The sense of urgency to sail at the earliest possible moment was exemplified by an internal memorandum sent to assistant managers by ... the operations manager at Zeebrugge. It is dated 18th August 1986 and the relevant part of it reads as follows:
>
> "There seems to be a general tendency of satisfaction if the ship has sailed two or three minutes early. Where a full load is present, then every effort has to be made to sail the ship 15 minutes earlier ... I expect to read from now onwards, especially where FE8 is concerned, that the ship left 15 minutes early ... put pressure on the first officer if you don't think he is moving fast enough. Have your load ready when the vessel is in and marshall your staff and machines to work efficiently. Let's put the record straight, sailing late out of Zeebrugge isn't on. It's 15 minutes early for us." (Department of Transport 1987: 11)

The Court found it difficult to understand why the loading officer could not remain on G Deck until the doors were closed before going to his station on the bridge, given that the operation to close the doors could have been completed in less than three minutes. Townsend had argued that the disaster could have been avoided if the Chief Officer had waited on deck for another three minutes—which was true, but the Court noted that the deck officers always felt under pressure to leave early and Townsend took no steps to ensure that the Chief Officer waited on deck until the doors were closed.

On the 6th of March 1987, the *Herald of Free Enterprise* was running late, ultimately sailing away five minute late.

At its heart, the *Herald of Free Enterprise* revolves around on-time running, a manifestation of production before safety that is known in transport-related industries.* All of the criticisms of management identified in the inquiry, and all of the errors by the ship's personnel on the day of the tragedy, are tainted to some extent by the spectre of "production before safety": in this case, the pressure to sail early.

Home insulation

Bureaucratic pressure to bring a program into existence quickly is a common theme of history and played out recently in the Australian home insulation program that was announced in February 2009. In response to the global economic crisis in late 2008, the Australian government announced a $42,000,000,000 national building and jobs plan as part of a series of economic stimulus measures. A key element of the play and was a $3.9 billion energy efficient homes package. This package was designed to generate economic stimulus in the housing and construction industry and was to improve the energy efficiency of Australian homes. Installing insulation in existing homes was regarded as one of the most cost-effective opportunities to improve residential energy efficiency. In light of these, the program included a home insulation program which provided incentives for homeowners and occupiers to have insulation installed. This program was costed at $2.8 billion over $2^1/_2$ years. The programs allow the homeowners and occupiers to be reimbursed up to $1600 (and landlords or tenants up to $1000) before having home insulation installed. For a period of the program running 31 weeks the installer was paid the rebate directly rather than the householder, which was designed to increase the number and the speed of payment of the claims for reimbursement.

Under the programs some 1.1 million roofs were insulated at a cost of $1.45 billion. However, the program was found to have a number of significant problems, including problems relating to the management of the health and safety risks arising from the project. One of the criticisms of the project was that there was no adequate control on the competence of installers of home insulation, in particular metal foil insulation, which led to unsafe installation resulting in house fires and the deaths of some installers—particularly young and inexperienced installers.

Legal proceedings commenced, with some companies being prosecuted in relation to unsafe work practices. In addition the program was audited by the Australian National Audit Office. The Audit Office identified that:

---

* See for example the useful discussion of the 1999 Glenbrook rail disaster in Hopkins, A. 2005. See also Cullen, L. 2001.

The program was developed in a very short period
of time between 3 February, 2009 and 30 June, 2009
as a stimulus measure to respond to the global
financial crisis.

...

In large measure, the focus by the department
on the stimulus objective overriding risk management practices that should have been expected
given the inherent programme risks.

...

There were insufficient measures to deliver
quality installations and, when the volume of issues
requiring attention by the department increased,
the department had neither the systems nor capacity to deal with this effectively. (ANAO: 2010: 27)

Deepwater Horizon

In the period leading up to the disaster on the Deepwater Horizon on 20
April 2010, the operations were under significant cost and schedule pressure. The Macondo well was nearly six weeks behind schedule and more
than $58 million over budget. The ongoing cost to BP for the lease of the
Deepwater Horizon was as much as $1 million per day (Graham et al.
2011: 2). In a very real sense time was money.

It is perhaps unsurprising then that the operations on the Deepwater
Horizon would be subject to scrutiny about the role that cost cutting might
have played in the disaster.

On 14 June 2010 the Committee on Energy and Commerce, a committee of the House of Representatives of the United States Congress, wrote
to the then chief executive officer of BP, Tony Hayward (Waxman and
Stupak 2010). The letter described a number of areas of concern and that
the Committee was seeking information as part of its investigation into
BP's role in the Deepwater Horizon disaster—areas of concern where it
appeared that BP had put cost savings ahead of safety:

The Committee's investigation is raising serious
questions about the decisions made by BP in the days
and hours before the explosion on the Deepwater
Horizon. On April 15, five days before the explosion,
BP's drilling engineer called Macondo a "nightmare
well." In spite of the well's difficulties, BP appears to
have made multiple decisions for economic reasons

that increased the danger of a catastrophic well failure. In several instances, these decisions appear to violate industry guidelines and were made despite warnings from BP's own personnel and its contractors. In effect, it appears that BP repeatedly chose risky procedures in order to reduce costs and save time and made minimal efforts to contain the added risk. (Waxman and Stupak 2010: 1)

In the end, the National Commission report identified seven decisions that were made that saved time where a "less risky" alternative was available. The decision not to wait for more centralizers was one of those decisions that saved time.

Decision-making processes at Macondo did not adequately ensure that personnel fully considered the risks created by time- and money-saving decisions. Whether purposeful or not, many of the decisions that BP, Halliburton, and Transocean made that increased the risk of the Macondo blowout clearly saved those companies significant time (and money).

There is nothing inherently wrong with choosing a less-costly or less-time-consuming alternative—as long as it is proven to be equally safe. The problem is that, at least in regard to BP's Macondo team, there appears to have been no formal system for ensuring that alternative procedures were in fact equally safe. None of BP's (or the other companies') decisions in Figure 4.10 appear to have been subject to a comprehensive and systematic risk-analysis, peer-review, or management of change process. The evidence now available does not show that the BP team members (or other companies' personnel) responsible for these decisions conducted any sort of formal analysis to assess the relative riskiness of available alternatives.

Corporations understandably encourage cost-saving and efficiency. But given the dangers of deepwater drilling, companies involved must have in place strict policies requiring rigorous analysis and proof that less-costly alternatives are in fact equally safe. If BP had any such policies in place, it does not appear that its Macondo team adhered to them. Unless companies create and enforce such policies, there is simply too great a risk that financial pressures will systematically bias decision-making in favor of time- and cost savings. (Graham et al. 2011: 125–126)

The National Commission did however comment that they *"cannot say"* whether anyone consciously chose a riskier option because it would save time or money.

The issue of conscious decision-making was also a feature of the evidence given by Tony Hayward before the House of Representatives on 14 June 2010, and he was at pains to emphasise that putting costs ahead of safety was not acceptable to BP:

If I found at any point that anyone in BP put cost ahead of safety, I would take action. (Hayward 2010: 113)

...

Mr Hayward: There is nothing I have seen in the evidence so far that suggests that anyone put costs ahead of safety. If there are, then we will take action. (Hayward 2010: 113)

...

Mr Hayward: I am not the drilling engineer, so I am not actually qualified to make those judgments. Better people than I were involved in those decisions in terms of the judgments that were taken. And if our investigation determines that at any time people put costs ahead of safety, then we will take action. (Hayward 2010: 113–114)

...

Mr Hayward: I think that is, you know, a cause for concern. I would like to understand the context in which it was sent. And as I have said a number of times, if there is any evidence that people put costs ahead of safety, then I will take action. (Hayward 2010: 124)

...

Mr Hayward: That is what our investigation will determine, and that is what it is going to do. And if there is, at any point, evidence to suggest that people put costs ahead of safety, then I will take action. (Hayward 2010: 150)

...

Mr Hayward: As I have said, we need to wait for the results of the investigation to conclude. If there is any evidence whatsoever that people put costs ahead of safety in this incident, then we will take action. (Hayward 2010: 155)

...

Mr Hayward: I don't want to be evasive, but I genuinely believe that until we have understood all of the things that contributed to this accident, it is not easy to say what I would say. If there is evidence that costs were put ahead of safety, I would be both deeply disturbed, and we would take action. (Hayward 2010: 186)

The notion that individuals consciously or knowingly make decisions to compromise safety in the face of production pressures in many ways misconstrues the nature of the issue. As the National Commission made clear, the fault did not lay in choosing less costly options—that is an inherent pressure in any competitive business environment. Rather, the fault lay in the failure to properly assess the risks arising from the less costly or time-saving options, and it seems that by April 2010 BP's risk assessment was skewed in favour of production.

It was apparent that there was a growth strategy in BP focussed on cost, "capital efficiency", and "margin quality", supported by programs of staff reduction and cost cutting, as well as incentive programs for employees to save the company money (Gale 2010: 24).

The culture of BP by the time of the Deepwater Horizon disaster was one of operating at the edge of risk, and a common language seemed to have developed in pursuit of this. For example, at the 4th Annual Strategic Decisions conference, BP's then head of Exploration & Production, Andy Inglis, spoke of BP operating at the edge of risk, taking major risks, managing big risks for commensurate rewards, and operating at the "right edge of the technical envelope" (Gale 2010: 26). And it was this language and culture that flowed through to the decisions being made in regard to the operations on the Deepwater Horizon.

In its letter dated 14 June 2010, the House of Representatives identified an e-mail following up on the decision by Guide not to wait to install the recommended number of centralizers.

An e-mail from Brett Cocales, BP's operations drilling engineer, indicates that Mr Guide's perspective prevailed. On April 16, he e-mailed Mr Morel:

> Even if the hole is perfectly straight, a straight piece of pipe even in tension will not seek the perfect center of the hole unless it has something to centralize it.
>
> But, who cares, it's done, end of story, will probably be fine and we'll get a good cement job. I would rather have to squeeze than get stuck ... So **Guide is**

> **right on the risk/reward equation**. (Waxman and
> Stupak 2010: 8) [emphasis added]

It is difficult not to have at least some sense that the risk appetite being
expressed at the most senior levels of BP did not influence the decision-
making throughout the organisation.

In a footnote, the National Commission Inquiry noted that:

> The industry and the international community also
> failed to adequately communicate lessons learned
> from the Montara blowout, which for ten weeks
> beginning on August 21, 2009 spewed between
> 400 and 1500 barrels per day of oil and gas into the
> Timor Sea approximately 150 miles off the north-
> west coast of Australia. David Borthwick, Report of
> the Montara Commission of Inquiry (The Montara
> Commission of Inquiry, Australia, June 2010), 5, 26.
> According to the Report of the Montara Commission
> of Inquiry, released on November 24, 2010, many of
> the technical and managerial causes of the Montara
> blowout track those at Macondo. (Graham et al.
> 2011: 327, footnote 171)

One of those areas where the two inquiries tracked one another was
the examination of production over safety.

Montara Inquiry

Like the Deepwater Horizon, in the leadup to the disaster on the West
Atlas, the project was running behind schedule and over budget. The
Montara Inquiry and the investigations into the Deepwater Horizon have
raised serious questions about a culture of production before safety that
seems to have existed in both cases prior to the incident. A particular feature of the cost issue was "rig time". In both events there has been an
emphasis placed on the high daily rate paid to engage drilling rigs to
carry out offshore work in the oil and gas industry. The implication of
the high cost of rig time was that people were acutely aware of the cost of
maintaining drilling rigs in service and that there was a natural tendency
to want to get the work done as quickly as possible because of this.

As with the National Commission of Inquiry into the Deepwater
Horizon, the Montara Inquiry also found a prevailing culture that compromised safety in the pursuit of production goals.

Although PTTEPAA insisted in its oral and written submissions to
this Inquiry that it did not cut corners or seek to minimise costs where this

might compromise safety or well integrity, this claim does not bear scrutiny. The prevailing philosophy revealed by PTTEPAA's actions appears to have been to get the job done without delay (Borthwick 2010: 11).

The tension between the two objectives of saving time and cost on the one hand and operating safely on the other was well illustrated in Montara. That there were time and pressure in this case was undeniable—and largely not denied.

Mr Howe QC: Sir, Mr Wilson has given evidence that there is a very natural tendency on the part of the people out on the rig to want to get the job done as quickly as possible; do you accept that?

Mr Duncan: Yes.

Mr Howe QC: It is a known phenomenon, isn't it?

Mr Duncan: Yes, I'm not denying that.

Mr Howe QC: It is just like breathing air—that's what they do; that's right, isn't it?

Mr Duncan: Yes.

Mr Howe QC: In fact, I think there is almost, if you like, a known phenomenon of wanting to go from pre-spud to production as quickly as possible, and that's a general phenomenon on the rig; people get sort of implicated in it and they make a positive effort to acquit themselves favourably in that regard—that's right, isn't it?

Mr Duncan: That's fair. (Duncan 2010: 1369)

The difficulties that manifested during the course of the Inquiry were, first, the lack of evidence demonstrated by management to show how safety was properly considered in the light of the time and cost pressures; in other words, how could management show that when they made a decision that resulted in a time or cost saving, how could they show that they were confident that safety would not be adversely affected?

A second issue was the extent to which managers relied on assumptions—essentially assumptions that everyone could be relied upon to do the right thing when it came to safety.

An example of the first difficulty was the decision to use pressure containing corrosion caps instead of a cemented barrier as part of the well suspension program. The original well suspension plan involved the use of cement barriers, but the plan was changed on relatively short notice to replace the cement barriers with removable pressure containing corrosion caps. Counsel assisting the Inquiry used the decision to replace the cement barriers with pressure containing corrosion caps as an example of how management had not properly, at least in the opinion of Counsel assisting the Inquiry, considered the risk implications of the decision that they were making.

Mr Howe QC: In relation to the fitness for purpose of PCCs, as I understand it, you were concerned to downplay the significance of any cost savings; is that right?

Mr Duncan: I told you that I didn't think the cost savings were that significant.

Mr Howe QC: But it seems as though the pursuit of those cost savings is really very much part of a broader approach that you have; namely, "If something is fit for purpose and it is lower cost, then that's the way we will go forward"?

Mr Duncan: That's fair.

Mr Howe QC: Do you accept that whilst something might be regarded as fit for purpose in one sense, it can, nonetheless, also raise other difficulties or issues that may need to be the subject of special management?

Mr Duncan: That's possible.

Mr Howe QC: I want to suggest that PCCs is a very, very good example of that proposition; do you agree with that?

Mr Duncan: Yes. Each item of equipment which you choose, fit for purpose, will have its own special set of circumstances. PCCs are not excluded from that.

Mr Howe QC: No. In fact, I'm suggesting that they are a good example of that.

Mr Duncan: Yes.

Mr Howe QC: Because, in one sense, they look cheaper and they look broadly equivalent in terms of their barrier protection function; that's right, isn't it?

Mr Duncan: Yes.

Mr Howe QC: But, in another sense, they can bring their own special risk implications for various reasons; would you agree?

Mr Duncan: Yes.

Mr Howe QC: If you are going to remove a cement plug, you are going to have the BOP over the rig at the time and drilling through and you will have your drilling fluid in place and so on; that's right, isn't it?

Mr Duncan: Yes.

Mr Howe QC: You would only get to that point of removing that cement plug when you are going to keep in place a level of well control; that's right, isn't it?

Mr Duncan: That's correct.

Mr Howe QC: Whereas here, for instance, in relation to the 9-5/8" PCC, it was readily removed, wasn't it?

Mr Duncan: Yes.

Mr Howe QC: And it wasn't reinstalled; that's right?

Mr Duncan: That's correct.

Mr Howe QC: Because it was a batch drilling exercise, attention was diverted to getting on to the other wells; that's right, isn't it?

Mr Duncan: Yes.

Mr Howe QC: And insufficient attention was given to the risk implications of bringing forward the removal of the PCC, without adequate well control, by some four days; you will agree?

Mr Duncan: With hindsight, yes.

Mr Howe QC: Had there been a cement plug and not a PCC, you will agree that the course of events might have panned out differently?

Mr Duncan: I agree.

Mr Howe QC: So what seemed like a great, fit-for-purpose, low-cost option, namely, substituting a cement plug for a PCC --

Mr Abbott: I'm sorry, "low" might be "lower".

Mr Howe QC: So you will agree that what looked at one point in time as a lower-cost, fit-for-purpose option did have implications that required a level of risk management and attention to detail that we now know wasn't given?

Mr Duncan: Yes. (Duncan 2010: 1391–1393)

This first difficulty—not being able to demonstrate how safety is taken into account when making changes that impact on time and cost—was compounded by fundamental assumptions that underpinned management's approach to managing safety risks. In effect, the evidence in the Montara Inquiry was that senior management did not turn their minds to safety risks on the drilling rig because it was assumed that everybody would do the right thing.

In the Montara Inquiry at least, it was not suggested that this was deliberate, but that the focus on cost could lead to compromised decisions with respect to safety.

Cost-cutting, or cost-effectiveness issues, and the roles and responsibility of management to understand the impact of cost-cutting decisions were raised squarely in the Montara Inquiry, with a clear line of questioning directed at senior managers involved in decision-making prior to the accident that questioned their prioritisation of safety: was production and cost-cutting put before safety?

Conclusion

In the context of production and safety, the commitment to safety is often expressed in the face of budget cuts by pronouncements that there will be no reduction in the organisation's health and safety budget. While no doubt well meaning, this again represents a misunderstanding of the issues at work. At a basic level it sends a message that health and safety is somehow independent of the business and can be quarantined as some sort of different business metric that will deliver good health and safety outcomes in the face of "business decisions" made to accommodate the budget pressures, business decisions that are not within the realm of the health and safety budget but can have significant decisions on health and safety outcomes, decisions such as hiring freezes, deferring training—albeit not health- and safety-specific training and deferring maintenance, or deferring new equipment purchases.

If you consider all of the cases examined in this chapter, there was nothing in the context of the production/safety balancing that would have been better managed or alleviated by a larger health and safety budget. Indeed, many of the factors that we have seen that contribute to accidents are not factors within a typical health and safety budget at all, and indeed are not things that are meaningfully addressed within the general framework of health and safety management.

References

ANAO (Australian National Audit Office). 2010. *Audit report No. 12 2010-11. 2010. Home Insulation Program.* ANAO. Canberra. http://www.anao.gov.au/uploads/documents/2010-11_Audit_Report_No_12.pdf (Accessed 21 February 2011).

Borthwick, D. 2010. *The report of the Montara Commission of Inquiry.* Montara Commission of Inquiry, Canberra. http://www.ret.gov.au/Department/Documents/MIR/Montara-Report.pdf. (accessed 25 November 2010).

Cullen, Lord. 2001. *The Ladbroke Grove rail inquiry.* Health and Safety Executive. London: HMSO.

Department of Transport.1987. *MV Herald of Free Enterprise report of Court No. 8074 Formal Investigation.* London. http://www.maib.gov.uk/publications/investigation_reports/herald_of_free_enterprise/herald_of_free_enterprise_report.cfm (accessed 23 November 2010).

Duncan, C. 2010. *Transcript: Montara Commission of Inquiry.* http://www.montarainquiry.gov.au/transcripts.html (accessed 29 September 2010).

Gale, W. E. 2010. Perspectives on changing safety culture and managing risk. *The Macondo Blowout.* 3rd Progress Report, Appendix C. Deepwater Horizon Study Group. http://ccrm.berkeley.edu/pdfs_papers/bea_pdfs/DHSG_ThirdProgressReportFinal.pdf (Accessed 1 January 2011).

Graham, B. et al. 2011. *Deep water: The Gulf oil disaster and the future of offshore drilling.* Report to the President. National Commission on the BP Deepwater Horizon Oil Spill and Offshore Drilling. http://www.oilspillcommission.gov/final-report (accessed 11 January 2011).

Guide, J. 2010 *Transcript: Deepwater Horizon Joint Investigation,* 22 July 2010: http://www.deepwaterinvestigation.com/go/doctype/3043/56779/.

Hayward, T. 2010. Transcript: U.S. House of Representatives, Subcommittee on Oversight and Investigations, Committee on Energy and Commerce. The role of BP in the Deepwater Horizon oil spill, 17 June 2010. Washington, D.C.   http://energycommerce.house.gov/documents/20100617/transcript.06.17.2010.oi.pdf (accessed 23 November 2010).

Hopkins, A. 2005. Safety, culture and risk: The organisational causes of disasters. Sydney: CCH.

Waxman, A., and B. Stupak. 2010. Letter from the U.S. House of Representatives subcommittee on Oversight and Investigations to Tony Hayward, Chief Executive Officer BP PLC dated 14 June 2010.

chapter 12

Managing the obligations

> "I didn't feel particularly strongly about it having to
> be reinstalled, and I didn't want to undermine the
> people who were running the rig at the time. I was
> out from town. I didn't want to come and overrule
> them all the time."*

Introduction

If there is a key or central theme to this book, it is assurance. Either as an
organisation or as an individual manager, how do we have assurance that
the critical health and safety risks in our business are identified, understood, and controlled?

How do we know that the policies and procedures are, and continue
to be, effective?

Are the rules the right rules to manage the risks, and are they followed correctly?

Do we recognise changes in our business, and do we understand their
potential impact on safety and health?

Do the people we entrust to carry on the business have the skills, experience, training, resources, and authority to meet the health and safety
expectations of the organisation?

Do we identify and understand warning signs? Do we respond appropriately to them?

Do we really understand the safety and health lessons from incidents—our own or those in other organisations—and do we learn those
lessons and implement them?

Are we genuinely prepared to sacrifice production for safety?

Are we truly prepared to stop the job?

What do you know?

If assurance is the key theme, then the other side of the coin is assumption, the readiness of management to assume that safety is being well
managed, simply because no one has told them otherwise.

---

* Duncan 2010: 1323

> It appears to me that there were significant flaws in the quality of [the] management of safety which affected the circumstances of the events of the disaster. Senior management were too easily satisfied that the [Permit to Work] system was being operated correctly, relying on the absence of any feedback or problems as indicating all was well. (Cullen 1990: 238)

And it is this prevalence of assumption, absence of assurance, and lack of challenge that seems to me to lie at the heart of management failure when it comes to safety and health.

Bata

We started this book by looking at the Bata decision.* In that decision the Court commented on the failure of managers to be aware of what was going on:

> As the "on-site" director, Mr Weston had a responsibility in this type of industry to personally inspect on a regular basis, i.e., "walkabout". To simply look at the site "not too closely" 20 times over his four-year tenure does not meet the mark.†

Montara

One of the critical decisions in the sequence of events leading to the blow out at Montara was the decision that was made to remove the 9⅝" PCCC ahead of schedule and then not replace it. It did not need to be removed when it was, and there was certainly nothing to prevent it being replaced. One of the reasons, it seems, that it was not replaced was a reluctance by management to interfere in the way that others went about their business:

Mr Howe OC: So if the possibility or prospect of down rig time, lost rig time, wasn't the primary factor, what was, that is, in your decision not to instruct that the 9-5/8" PCC be reinstalled?

Mr Duncan: I didn't feel particularly strongly about it having to be reinstalled, and I didn't want to undermine the people who were running

---

* R v Bata Industries Ltd 7 C.E.L.R. (N.S.) 245, 9 O.R. (3d) 329 (**Bata**)
† ibid, [173]

the rig at the time. I was out from town. I didn't want to come and overrule them all the time. (Duncan 2010: 1323)

Another issue that arose in the inquiry was the presence of "bubbles" at the top of the well when the 9⅝" PCCC was removed. Mr Duncan was asked about the presence of the "bubbles":

Mr Howe OC: Who observed those bubbles?

Mr Duncan: It would have been Brian plus a couple of Seadrill personnel. I don't know their names.

Mr Howe OC: Did you have a personal conversation with Brian about what, in fact, he had observed?

Mr Duncan: I have now, yes.

Mr Howe OC: No, I mean on 20 August.

Mr Duncan: No.

Mr Howe OC: So what information did you have about these bubbles, when you were on the rig on 20 August, at the time or shortly after removal of the 9-5/8" PCC?

Mr Duncan: I don't think I spoke to Brian. I might have spoken to Paul.

Mr Howe OC: And Paul might have spoken to Brian or might have spoken to someone who spoke to Brian; is that right?

Mr Duncan: Yes.

Mr Howe OC: So you were content to rely on a hearsay—perhaps hearsay upon hearsay—account of the bubbles that were observed at the time of removal of the 9-5/8" PCC; is that right?

Mr Duncan: The bubbles were reported prior to removal of the 9-5/8" PCC. Yes, I was prepared to rely on what people told me. (Duncan 2010: 1317)

Whether these failures could be described as any way causally connected to the eventual incident depends, I think, on your perspective, but the attitude (whatever view you take of it) certainly does speak to the issue of how far managers ought to be going to be positively satisfied that what is happening in their area of responsibility is safe.

In isolation, a single incidence of a manager's unwillingness to "tell people what to do" and a reliance on hearsay might not tell us much about management's attitudes to checking safety—but this is not isolated, and is seems to be the rule rather than the exception.

Deepwater Horizon

When VIPs from BP and Transocean were on the Deepwater Horizon, they were there in large part it seems to advocate safety. They were doing the right thing. Yet the reluctance to challenge and seek assurance in relation to safety seems to have infected the visit in a way that undermined their activities.

When the VIPs were doing their tour, they came across workers who were having difficulty with the critical negative pressure tests:

A: I believe his words were, "We were having a little trouble getting lined up, but it's no big deal."

                              ...

Q: And that was a conversation about the negative test?
A: That was the end of the conversation with respect to negative test. [*]

There was a recognition that assistance may have been required, and it seems that the expertise was available amongst the VIPs to assist, but they moved on, not wanting to *"distract"* people:

> I said, "Hey, how's it going, Dewey? You got every-thing under control here?"
>
> And he said, "Yes, sir."
>
> And there seemed to be a discussion going on about some pressures or a negative test. And I said to Jimmy and Randy Ezell, "Looks like they're hav-ing a discussion here. Maybe you could give them some assistance." And they happily agreed to do that.[†]

                              ...

> When we left the meeting -- or prior to the meeting, if I can go back a little bit. I haven't got my statement with me to -- I'd asked Jimmy, I said, "Everything all right up on the rig floor there? Get everything sorted out?"
>
> He gave me a thumbs up and said everything was okay.[‡]

                              ...

> There might have been five or six, seven indi-viduals, and it appeared to me that they were having

[*] Transcript: Deepwater Horizon Joint Investigation, 26 August 2010, 136. http://www.deep-waterinvestigation.com/go/doctype/3043/56779/. (accessed 20 November 2010)
[†] Transcript: Deepwater Horizon Joint Investigation, 23 August 2010, 443. http://www.deep-waterinvestigation.com/go/doctype/3043/56779/. (accessed 20 November 2010)
[‡] ibid, 445.

> discussion about operations, and I didn't feel that it
> was a good time for a bunch of tour group to be, you
> know, involved with the operation or distract them.
> And I thought that Jimmy Harrell and 12 Randy
> Ezell, who were on the tour with us, you know,
> because it was about drilling activities or what they
> were doing, the negative test, there was some pres-
> sures didn't seem right, that they would be the best
> individuals to stay there.*

It has to be asked whether or not there should have been—or if there is a reasonable expectation that there should be—further challenge, questioning, and assurance? Given what eventually occurred, it is easy to say yes, but challenge and assurance can also boarder on micro-management, and how do we know when the line is crossed?

Q: You were the wells director, and you are currently the BP for drilling and completions operations. But in those roles, how did you ensure that the people that are -- that are answering to you are actually doing their job if you're not doing spot-checks or having some type of accountability to make sure they're doing what you're paying them to do?

A: We would -- we would check with people what they're doing, but this would go down through the chain of command. So, you know, I -- I wouldn't necessarily go direct to a single person, I may go to his manager. Are we on track? Are things going okay? Are we managing the way we should be? The way we're set up.†

Managers being prepared to rely on what people tell them, relying on the "thumbs up", and checking whether things are "okay".

How do you know that your critical risks are being controlled?

Repeated failures

Of course the tragedy of the position that we find ourselves in after the most recent disaster of the Deepwater Horizon is that there are no new lessons, just repeated failures followed by recycled rhetoric and outrage.

The failures of management on the Deepwater Horizon were the same failures identified in the Montara Inquiry and the Piper Alpha Inquiry, the same failures at Longford in Victoria and at Texas City, at Chernobyl

---

* ibid, 457.

† Transcript: Deepwater Horizon Joint Investigation, 25 August 2010, 89. http://www.deepwaterinvestigation.com/go/doctype/3043/56779/. (accessed 20 November 2010)

and in case of the *Herald of Free Enterprise*. They are the same failures of management we see in any catastrophic failure that you care to examine, and they are failures that underpin every workplace incident, no matter how minor:

> How do you know that the risks in your business,
> in your area of responsibility, are being managed?

One of the things that has struck me particularly in observing accident inquiries and court cases over the last 20 years is the inability of managers, particularly senior managers, to be able to provide any meaningful positive position statement on safety. The position is always clichéd: *'I take safety seriously'*, *'Nothing is more important than safety'*, *'There is nothing so important that we can't do it safely'* and of course, *'Everyone has the right to stop the job.'*

In all of my research and reading I have yet to see a senior manager in a major accident inquiry be able to clearly state:

> These are the critical risks that I was responsible
> for ... and this is how I know that they were being
> controlled.

Managers know when production is down. They know when schedules are overrun and they know when budgets are blown. Why don't they know when risks aren't being controlled?

Managing expectations

While I have no doubt that this sounds like a criticism of managers, it is not intended to be. I think that it is too simplistic and potentially dangerous to say that major accidents—or any accidents—are just a result of sloppy management. After all, as the National Commission of Inquiry into the Deepwater Horizon identified: *"complex systems almost always fail in complex ways"* (Graham et al. 2010: viii).

Having worked as a safety manager in a complex organisation, I can quite happily say that maintaining the line of sight to critical risks in a business is hard, even overwhelming at times.

While we would all like to work in organisations with ideal safety cultures that allowed everyone the time and space to appreciate and manage all safety risks appropriately, that is simply not the reality for most organisations. To suggest that safety culture and the development of good safety cultures are the key to the prevention of accidents (while aspirationally true) denies the here-and-now reality for most organisations and provides precious little guidance to managers who may not get to make the call about decisions affecting safety and safety culture—they just have to live with them.

To me, a striking feature of all the cases discussed in this book (and I dare say any accident that is given a reasonable level of scrutiny) is the commonality, even universality of the management failures.

What risks?

Of course, all this brings up a number of questions about risk. What should I be focussed on? In most mid- to large-sized organisations the volume of safety-related processes, policies, and procedures is enormous, and in some cases both overwhelming and potentially detrimental to safety, as the Baker Panel review identified in its discussion on "initiative overload".

Surely it is not reasonable to expect a manager to know every safety-related risk and associated process.

If we did not appreciate it before, the Texas City Refinery explosion was explicit: focussing on personal safety—slips, trips and falls—does not prevent major accidents. This lesson was clearly underscored by the events on the Deepwater Horizon, when senior managers were on board the rig sharing safety lessons; but they were safety lessons only about slip and trip hazards and the integrity of harnesses for working at heights, and the managers missed the opportunity when it presented itself to hold discussions with workers and share lessons and experiences about problems that had the potential for, and it turns out that did have, catastrophic consequences.

Personal injury risk dominates the safety landscape. Companies measure it. Industries and industry associations compare member company performance by it. Safety regulators measure year on year performance by it. Workers' compensation and other insurance costs are driven by it.

As a manager it is hard to avoid it.

But ask yourself this question: How many of the companies that were looked at in the cases in this book and how many individual managers had to answer difficult, potentially career-destroying questions about personal safety issues? It seems none.

This is a book about management obligations. Individual accountability.

When we look at the cases, we can see four principles that underpin what I have described as the minimum expectations of managers:

1. That managers understand their obligations for safety and health and the critical risks in the business—particularly those that they are responsible for
2. That managers have processes in place to bring any failures in the safety management processes to their attention; that they know when things are going wrong

3. That managers respond personally and in a timely manner to any concerns about safety
4. That from time to time managers personally verify that safety is being managed in accordance with their expectations; that what they believe to be happening is actually happening

But here is the reality. When it really matters, when managers get called to account for their actions, is when people die. So what are the risks in your business that could kill people and how do you know that they are controlled?

Not what you assume.
Not what you have been told.
Not what you think.
What do you know?

Moura

At approximately 11:35 pm on 7 August 1994, an explosion occurred in the Moura No. 2 underground coal mine in Queensland, Australia.

There were twenty-one people working underground at the time. Ten men from one area of the mine escaped within thirty minutes of the explosion, but eleven men from another area failed to return to the surface.

A second, more violent explosion occurred at 12:20 am on 9 August 1994. Following this second explosion, rescue and recovery attempts were abandoned, and the mine was sealed at the surface.

The Inquiry into the incident found that the first explosion resulted from a failure to recognise, and effectively treat, a heating of coal in one of the panels. This, in turn, ignited methane gas which had accumulated within the panel after it was sealed. The Inquiry did not reach a finding regarding the cause of the second explosion.

Prior to the explosion, before the start of the night shift on 7 August 1994, Moura mine management were aware of potentially dangerous gases that may have been underground. There were three key questions facing the mine management on the night of 7 August 1994:

- Should the workers go underground?
- What should the workers be told?
- What should management do if the workers raised concerns about the atmosphere in the mine passing into the explosive range?

On the afternoon of 7 August, Mr Squires, the shift undermanager, raised the question with Mason, the undermanager in charge, concerning an appropriate course of action if the workers had any concerns about the atmosphere passing into the explosive range. Mr Mason's response was that if no one else raised a concern, then neither should Mr Squires.

The following passage is from the evidence of Mason in relation to his conversation with Squires on the Sunday evening:

Q: Can we move on with the conversation? Did he ask you something about how he should deal with the matter at the start of shift that night?

A: Yes, he did.

Q: What did he say about it and what did you say about it?

A: Well, I was rather confused. Michael asked me then at that point, after I had explained to him my -- how -- I'd explained to Michael how I would deal with the situation, then he asked me how I wanted him to deal with it, how I wanted him to broach the subject at start of shift. I was somewhat confused, because I had just spent time going through that with him.

Q: Again, if you could try to give the conversation that took place as best you can remember on this aspect of the telephone call?

A: Michael asked me how I wanted him to approach the subject at the start of shift; did I want him to summon all the men together and give them a run-down of the events that had transpired? I told Michael there was no need to do that. I did not believe there was a need to do that, as quite a number of the people who worked permanent night shift had been involved with those events on the weekend. They had been there through the sealing, there were deputies that had been at work through the sealing and the shifts preceding and the shifts sub-sequent. I told him that I thought that the men would be well aware of the situation as it was.

Q: But when you say "the situation as it was", what are you referring to there?

A: Well, that course of events that had transpired over the weekend.

Q: But I'm just interested in what you mean when you say the men would have been well aware of what the situation was. Can you explain that reference—"the situation as it was"?

A: Well, all those things that we have spoken about up to ...

Q: Just run through them?

A: That the panel had been sealed as a precautionary measure as a result of a number of observations that had been made—I guess basically that's it.

Q: The men that were to go down on the night shift that night, do you say that they would have been aware of this report ... about a slight tarry smell on the Friday afternoon?

A: I believe they would have been yes.

Q: How would they have become aware of that?

A: The people who were involved on the sealing process had that -- had those circumstances explained to them.

Q: But not all of these men that were to go down on the Sunday night had been involved in the sealing process, had they?

A: That's correct.

Q: So, on what basis did you expect that those people would have become aware of this report ... on the Friday afternoon of a slight tarry smell?

A: News around the mine—there is quite a good grapevine at work. People always seem to have knowledge of events that transpire in the mine.

Q: So, you were relying on the grapevine, in effect; is that what you are saying?

A: Yes. (Windridge 1995: 44–45)

How do you know that the critical health and safety risks in your business are being effectively controlled?

You do not have the luxury of assumption.

References

Duncan, C. 2010. *Transcript: Montara Commission of Inquiry.* http://www.montarainquiry.gov.au/transcripts.html (accessed 29 September 2010).

Graham, B. et al. 2011. *Deep water: The Gulf oil disaster and the future of offshore drilling.* Report to the President. National Commission on the BP Deepwater Horizon Oil Spill and Offshore Drilling. http://www.oilspillcommission.gov/final-report (accessed 11 January 2011).

Windridge, F. W. 1995. Report on an accident at Moura No. 2 underground mine on Sunday, 7 August 1994. Wardens Court, Queensland.

Index